'Student ambassadors have been a central part of recent initiatives to inspire more young people to take up careers in science, engineering and the other STEM subjects, but there is surprisingly little research into the dynamics of pupil ambassador relationships. This book is based on detailed research into a number of student ambassador schemes and comes up with some unexpected insights about the learning that occurs. It is essential reading for anyone interested in promoting STEM among young people.'

Professor Matthew Harrison, Director, Engineering and Education at the Royal Academy of Engineering

'This thoughtful and accessible book provides a much-needed critical approach to understanding the current, and potential, role of student ambassadors. A "must read" for anyone ⋯ ˙ ˙ participation fiel

Professor Louise Archer, Department of Educ King's College Lon

'This book makes an important contribution to debates about widening participation. By drawing on detailed empirical work in two higher education institutions, it presents fascinating evidence about the role of student ambassadors, and the impact they have on pupils' learning identities.'

Professor Rachel Brooks, Department of Sociology, University of Surrey

'Clare Gartland's excellent contribution to research on widening participation in higher education illuminates the important dimension of student ambassador work. *STEM Strategies* provides both a macro-level analysis of policy and globalized discourses and a close-up, detailed study of the relationship between ambassadors and students. It is an important book for the development of widening participation strategy, policy and practice and for reaching a deeper understanding of equity issues in higher education.'

Professor Penny Jane Burke, School of Education, University of Roehampton

STEM Strategies

STEM Strategies
Student ambassadors and equality in higher education

Clare Gartland

A Trentham Book
Institute of Education Press

First published in 2014 by the Institute of Education Press, University of London, 20 Bedford Way, London WC1H 0AL

ioepress.co.uk

© Clare Gartland 2014

British Library Cataloguing in Publication Data:
A catalogue record for this publication is available from the British Library

ISBNs
978-1-85856-617-7 (paperback)
978-1-85856-618-4 (PDF eBook)
978-1-85856-619-1 (ePub eBook)
978-1-85856-620-7 (Kindle eBook)

All rights reserved. No part of this publication may be reproduced, stored in a retrieval system, or transmitted in any form or by any means, electronic, mechanical, photocopying, recording or otherwise, without the prior permission of the copyright owner.

Every effort has been made to trace copyright holders and to obtain their permission for the use of copyright material. The publisher apologizes for any errors or omissions and would be grateful if notified of any corrections that should be incorporated in future reprints or editions of this book.

The opinions expressed in this publication are those of the author and do not necessarily reflect the views of the Institute of Education, University of London.

Typeset by Quadrant Infotech (India) Pvt Ltd
Printed by CPI Group (UK) Ltd, Croydon, CR0 4YY

Cover image: Ventilation and escape tunnel on the Jubilee Line, London Underground.
© QA Photos/Jim Byrne

Contents

Glossary of terms	viii
Acknowledgements	ix
About the author	ix
Foreword	x
Introduction	xi
1. Widening participation and the landscape of policy and research	1
2. Widening participation and STEM: the cases of engineering and medicine	14
3. Student ambassadors and mentoring	26
4. Analysing policy and practice: a multi-stranded approach	40
5. Meanings of marketing	57
6. Learning practices and identities	88
7. Social relationships and identities	115
8. Assumptions, practices and potential	148
Appendix	160
References	164
Index	181

Glossary of terms

Aimhigher	a programme funded by BIS, HEFCE, the Skills Funding Agency and the Department of Health with the ambition to widen participation in HE
AEP	Accessing Engineering Project
ATHENA	a network for women in business
BIS	Department for Business, Innovation and Skills
D&T	design and technology
EAL	English as an additional language
FE	further education
G&T	gifted and talented
GCSE	General Certificate of Secondary Education
GNVQ	General National Vocational Qualification
HE	higher education
HEFCE	Higher Education Funding Council for England
HEIs	higher education institutions
HESA	Higher Education Statistics Agency
IT	information technology
NAE	National Academy of Engineering
RAEng	Royal Academy of Engineering
SIVS	strategically important and vulnerable subjects
SET	science, engineering and technology
STEM	science, technology, engineering and maths including medicine
UKRC	UK Resource Centre for women in science, engineering and technology
WiC	Women in Computing
WISE	Women into Science and Engineering (works with industry and education to inspire girls and attract them into STEM studies and careers)
WP	widening participation
…	omission in transcriptions (one or more words, or an extract)

Acknowledgements

Many thanks to the project organizers, student ambassadors and school students who gave up their time to talk to me and made this study possible. Thanks also to Professor Matthew Harrison and Professor Miriam David for their support, expertise and advice, and to Anna Paczuska for being a constant source of support and encouragement.

About the author

Clare Gartland started her career working in sixth-form colleges in Leicester and south east London as a teacher and middle manager. She then moved into teacher training, working as a senior lecturer for Canterbury Christ Church University. Having developed an interest in widening participation during a fellowship at South Bank University in 2003, Clare increasingly moved into working as an independent researcher and evaluator of projects, aiming to develop ways to support pupil engagement and progression to higher education, particularly in science, technology, engineering and maths (STEM). For the past ten years, she has worked closely with colleagues in universities and at the Royal Academy of Engineering on the development of outreach programmes, engaging with schools, engineering employers and organizations such as the Science and Learning Centres, STEMNET and Aimhigher. Having completed her PhD with the Institute of Education in 2012, she is now continuing her research based at University Campus Suffolk, where she leads the MA programme in learning and teaching.

Foreword

Widening participation and access to higher education in the United Kingdom and abroad has become a hot topic in the twenty-first century. It is mainly about improving chances for the children of socially and economically disadvantaged families, ensuring that these become 'first-in-the-family' to go to university. Equality and social justice lie behind these policy and practice mantras. Curiously there are few studies on the now increasingly important subjects of STEM – science, technology, engineering and medicine or mathematics – to the future of our knowledge economy. Even more curiously there are no studies of how students themselves feel about being encouraged to study these vital subjects at university.

Clare Gartland has developed a unique study of how two contrasting universities in inner London strove to encourage young people into engineering and medicine, using a scheme created for just that purpose – student ambassadors. She has used her strong pedagogic skills and insights to explore how the young people she met as student ambassadors and potential university recruits felt about these interactions and how their own identities and feelings were being transformed in the processes of learning about the excitement of engineering or medicine. She has also ably dissected the very notion of ambassadors as part of a growing process in the business of higher education. Her forensic skills are ably deployed to ask questions about these new policies and practices and how they might impact upon the future of higher education in an increasingly global and marketized economy. Her significant achievement is to draw upon the voices of several diverse groups of students to both deepen and enliven her broader arguments about the creative potential of student ambassadors to STEM strategies.

This is an important, indeed essential, contribution to our understanding, and making, of a more equal higher education.

Professor Miriam E. David
Institute of Education, University of London

Introduction

Ten years ago, I carried out a small study for a local education authority in south London exploring young people's experiences of transition points in their secondary education. One story from this study has relevance to the focus of this book and contributed to my own interest in the possibilities presented by the work of ambassadors. Richard, a 15-year-old student from a white working-class background, was studying for his General Certificates in Secondary Education (GCSEs) and a General National Vocational Qualification (GNVQ) in Information Technology (IT) at our first meeting. His father was a local mechanic and his family had always lived in this area of London, but, unlike many of the students I spoke to, his ideas for the future were not constrained by this background. He talked in detail of his plans to work in Canada, to be the first in his immediate family to go to college to study post-16 and to be the first in his entire extended family to go to university.

It became clear during my interviews with him that various adults outside his immediate family and social networks had played significant parts in his course choices and life plans. He spent most of his free time at school in the print room, and this had enabled him to build close relationships with technicians and the IT network managers. These relationships seemed to be responsible for shaping Richard's ambitions and facilitating his creation of a new identity for himself. This identity was, in career terms, very different to that of other members of his family. Another important connection was a web developer at the local City Learning Centre, whom Richard met while he was on a course there. It seemed to be from this experience that Richard had, as he put it, 'got on the pathway with web development … he showed me examples of his work – with what he done at university'. Another connection for Richard was a network manager at the school, who had been to the same further education (FE) college to which Richard planned to go, and lived in Vancouver, where Richard said he also wanted to live.

It was these connections that had broadened Richard's 'horizons of action' (Hodkinson *et al.*, 1996) to the extent that, despite coming from a family who were rooted in south London, they were surprisingly 'global' (Ball *et al.*, 2000). Richard progressed to the FE college he had planned to go to and studied for an advanced GNVQ in which he gained a distinction. He then got a well-paid job in IT through the networks he had made at school. Two friends that he had made at his FE college progressed to an elite university in another part of the country on completion of their courses, and, after working for a year, he too obtained a place there. It was a Blairite success story. Richard successfully aspired to and moved physically into the middle class.

This story and others like it appear to present a simple solution to ameliorating entrenched social divisions in the UK, a solution that has been widely seized upon. Mentoring and similar schemes involving adults working with young people have increased exponentially in recent years in the UK and elsewhere (Colley, 2003), and are widely assumed to have a range of benefits for individuals and for society as a whole.

As part of a drive to increase and widen participation in higher education (HE) under New Labour (1997–2010), undergraduate students, or student ambassadors as they are commonly labelled, were employed by universities across the UK to work with school pupils to raise their aspirations and to support attainment. This picture is replicated in other developed countries; for example, similar schemes are described in higher education institutions (HEIs) in Australia, where government support for widening participation in HE has also been strong (Gale and Tranter, 2011), and in the US, where university students are extensively used by HEIs in STEM outreach. In the UK, schemes specifically designed for widening participation in STEM subjects were also supported by government, including the National HE STEM Programme; such schemes utilized undergraduate students as ambassadors to work with school pupils to encourage progression in STEM subjects. Despite recent funding and policy changes, student ambassadors are now an established part of the UK university landscape.

The practice of using 'ambassadors' or volunteers to work with young people has not been limited to universities. Such practices have been taken up by companies keen to increase the supply of qualified young people in the STEM pipeline. These companies are keen to widen and increase participation in STEM to ensure there are sufficient numbers of qualified young people in the field in future. In the UK, STEMNET, an organization that facilitates the work of STEM ambassadors in schools, now works with 2,500 separate employers (Straw *et al.*, 2011). Currently, 28,000 STEM ambassadors are working in various capacities with school pupils.

The work of ambassadors from industry is not limited to work through STEMNET. Individual companies are also running their own schemes in order to recruit. Anecdotal evidence indicates that these schemes are reaping rewards. The head of training at one global company recently described how

Introduction

using apprentices as STEM ambassadors in schools had contributed to a huge increase in applicants for apprenticeship schemes at the company:

> For a kid to be inspired, they need to talk to somebody who is two to three years older than them ... We have about 150 apprentices a year and we now attract about 4,000 applications a year.
>
> Manager, Skills and Business Learning Governance, Siemens plc

Given the extraordinary expansion of ambassador schemes in the UK and internationally, the successes claimed for them and the widespread assumption that ambassador work is a positive influence on the young people targeted, it is timely to reflect on lessons learned from New Labour's widening participation (WP) schemes and to consider what student ambassadors might contribute to promoting equality in HE in STEM subject areas.

Richard's story is clearly a success story. He progressed via a vocational route from a working-class background into a top university populated predominantly by middle-class students, many from private schools. But it is important to think critically about this success. While it may lead us to conclude that providing young people with role models will help them in similar ways, we should ask if this is an automatic outcome. What was it about the young people in Richard's experience that helped and inspired him? What aspects of their identity were significant to this process? Was it their gender or subject identities or both? Were other aspects of their identities significant? Was the physical space in which they were located important? Did the nature of the learning activities they undertook together contribute?

It is also important to draw away from this micro-level analysis of what works and to consider carefully which young people benefit from WP schemes promoting HE, how they benefit and whether the schemes are helping those who are most marginalized. Richard's story fits comfortably into individualized, meritocratic discourses that presume there are few working-class young people with the ability to access elite universities. The question that arises, then, is what happens to the other young people who, for whatever reason, do not manage to achieve this kind of success? Are they inevitably positioned through these discourses as failures, lacking the ability and ambition that such success demands? It is also important to move away from the current preoccupation with the benefits of HE for individuals and the private good, and to consider what promoting equality in HE, particularly in STEM subject areas, contributes to the public good.

This book aims to address these questions through a detailed study of WP interventions at two very different HEIs. I explore the institutions' practices of using student ambassadors in WP outreach work in STEM subjects. I particularly consider WP activities in engineering and medicine, two very differently positioned subject areas within STEM. Using a multidisciplinary approach, I provide a detailed account of these practices, exploring wider policy discourses that shape them, and providing critical and practical insights into what ambassadors can contribute to promoting equality in STEM subject areas and who this benefits. But this book is not simply a reflection on the UK New Labour's WP policy; it also poses questions about wider issues of social justice, diversity, inclusivity and equality in particular subject areas and career trajectories.

The first three chapters provide the contextual and background information that sets the scene for the empirical study that is the focus of the rest of the book. Chapter 1 provides a brief account of UK and international developments in widening participation to higher education. I consider the history of HE in the UK, how policy to increase and widen participation has evolved since the early 1960s, and how WP policy reflects approaches in other developed and developing countries in the race to upskill the workforce. Education is big business in itself and the HE system has been heavily marketized. The chapter outlines the stratification of the HE system in the UK and explores critiques of WP policy. I also consider the dearth of nuanced research exploring the relationship between pedagogy, learning and identity, an approach that has been neglected in the research and evaluation of WP initiatives in favour of a marked preoccupation with 'impact' or 'what works'.

Chapter 2 is focused on issues relating specifically to medicine and engineering. I consider the motivation behind outreach work in engineering and medicine and the issues affecting these subject areas. I outline some of the debates surrounding pupil engagement with STEM subjects and science more generally, and consider why there is such an entrenched gender divide.

In Chapter 3 I consider the assumption which underlies the policy discourse around ambassadors, namely that ambassadors are 'role models' for school pupils. The limited body of research into the work of student ambassadors is explored along with the wider international literature about mentoring schemes more generally.

Chapter 4 provides an account of the theoretical framework and outlines the multi-stranded approach that was developed to address the particular issues posed by exploring the work of student ambassadors with school pupils. The dearth of research into student ambassadors can be attributed partly to the challenges that the fleeting exchanges between ambassadors and school pupils pose in research terms. It is hoped that the approach outlined here provides a way forward for researching these relationships that moves beyond notions of impact to a nuanced understanding of learning processes. I also provide details of the institutions and participants in my research and outline the policy contexts within which activities were located.

Chapters 5 to 7 present the findings of the study. Chapter 5 explores the dominance of marketing discourses within the HEIs and in ambassador interactions with school pupils. Associated discourses of 'charity', 'deficit' and 'employability' that were embedded in the practices of institutions and ambassadors are also considered. I reflect on how this affects ambassador work and iterates with dominant neo-liberal discourses. In Chapter 6 I explore the ways in which ambassadors were positioned in different learning contexts and how this influenced their relationships with pupils and pupils' learning. I discuss the formal and informal attributes of these different learning contexts and how they affect interpersonal relations and learning in ambassador–pupil relationships. Chapter 7 focuses on processes of dis/identification. The iterations between learning contexts and intersecting aspects of ambassadors' and pupils' identities are explored and the complexity of these processes demonstrated.

Chapter 8 summarizes the findings of my study and the issues they raise. I discuss their implications for both practitioners and policy makers in widening participation, for academics and for those working in increasing participation in STEM subject areas.

Chapter 1

Widening participation and the landscape of policy and research

Interactions between pupils and student ambassadors are the subject of this book. Such interactions are at the centre of much work to widen participation in HE in the UK and elsewhere in the drive to upskill workforces internationally. Widening participation work in the UK has changed over the past two decades. The widening access movement, which emerged in the 1970s and 1980s out of liberal education initiatives promoting access to university, mainly for mature and adult students, grew under successive governments from the 1990s onwards and morphed into national government policy imperatives to widen participation and access. These policies aimed to raise aspirations and promote HE participation among young students as well as adults. Government imperatives dictated a focus on statistics and quantitative outcomes to demonstrate the success of schemes, the assumption being that quantitative methods ensure objectivity and identify most effectively what works (Hodkinson, 2004). While these approaches provide the statistics required for government reports and policy documents, they overlook the operation of the complex learning and social activities and processes which give rise to these statistics. More worryingly, these approaches promote assumptions about cause and effect in education policy, but the learning processes and social interactions that are behind these figures are not simple; more input does not necessarily lead to greater outcomes.

This cause and effect logic is dominant in policy development. The Conservative–Liberal Democrat Coalition Government's white paper *Higher Education: Students at the heart of the system* (2011) emphasizes the importance of 'wider availability' of information to pupils' ability to make informed HE choices. Research into student choice, however, demonstrates that it is unlikely that 'wider availability' of information will positively influence and inform young people's decision-making. The provision of more information does not necessarily make young people more informed. To find out what kind of information is effective in supporting student choice, it is important to consider not only the resources allocated by government policy and the impact of these resources on groups targeted by these policies, but also the social activities, learning processes and identities that develop in the context of policy initiatives. The research for this book strongly indicates that both learning processes and complex issues of identity are central to outcomes of these policy initiatives.

A brief history of widening participation and HE

The expansion of HE, often referred to as the global knowledge economy, is an international phenomenon. The Organisation for Economic Co-operation and Development declares that a 'well educated population' is 'essential for the social and economic well being of countries' (OECD, 2008: 30). Naidoo (2003: 250) observes that across the world 'intellectual capital' is viewed as 'one of the most important determiners of economic success'. There has been a 'massive expansion in HE internationally as governments move from an elite to a mass system' (Gale and Tranter, 2011: 31). The scale of the expansion of HE globally in recent years has been striking: there has been a 53 per cent increase in tertiary education globally since the year 2000 and, in China, an increase from 1.58 million participants in 1990 to over 23 million in 2006 (Liu, 2013). It is predicted that demand for HE will continue to grow from 97 million students worldwide in 2000 to over 262 million by 2025 (Morley, 2012).

Until the change of government in 2010, the approach in the UK was 'emphatic' about committing to increasing participation (Parry, 2011: 142). Successive New Labour governments (1997–2010) supported a raft of policy, strategy and funding measures to further these aims. As in other countries, the Labour administrations' concerns to widen participation reflected economic as well as social justice and equity issues. A paper from the Cabinet Office focused on 'social mobility' and the dual need for 'ensuring there are better jobs' and ensuring 'that people have fairer chances to access these jobs' (Cabinet Office, 2008: 12). Prime Minister Gordon Brown announced that 'in a globally competitive national economy, there will be almost no limits to aspirations for upward mobility' in the UK (Brown, 2008).

There are, though, critiques of the view that university is a universal good. Questions have been raised about the nature of the workforce that will be needed in twenty-first-century Britain. Morley suggests employability discourses are prone to overlook the possibility that many employers will still want a significant proportion of their workforce to occupy low-skilled jobs (Morley, 2001) and points out the increased numbers of students across the world who are accessing HE as having consolidated 'social status and the avoidance of downward social mobility rather than its extension' (Morley,

2012: 354). Steele (2000) warns of an increasingly divided UK society with large numbers of people involved in a new kind of work catering to the needs of those in high-paid employment. Brown *et al.* (2008) warn that globalized companies may well locate in developing countries with skilled workers where production is cheap. The need for upskilling the UK workforce may not be as clear as government discourses during this time suggest. However, the policy in the UK under Labour focused on upskilling the workforce and this involved a drive to expand uptake in HE.

If we consult the history of the HE system in the UK, it is clear that the economic driver that motivated the focus on increasing HE participation is not new. The economic needs of the country have driven the development of the HE system for several decades, reflecting increases in the economy's demand for more qualified workers. But the UK HE system has in many ways perpetuated rather than challenged social divisions. In the early 1960s, less than 10 per cent of the population, mostly white middle-class men, went to university (David *et al.*, 2010); the government's response to the Robbins Committee report of 1963 changed this dramatically. A binary policy was introduced with extended opportunities to access university for students on academic routes, through the development of new universities and the formation of polytechnics to cater for those on technological or vocational trajectories. As David *et al.* (2010: 8) state, 'a system of structured higher educational opportunities' that was based around different types of courses that linked to socio-economic status was 'thus embedded within UK policies and practices for expanding higher education from their inception'. In the 1980s, under a Conservative government, there were attempts to increase opportunities for mature and also for female students. In 1992, the binary policy was abolished and polytechnics became new universities.

Post-1992, the HE sector divided along structural lines. The term 'elite universities' generally refers to the traditional universities, some of which date from medieval times. Royal, one of the HEIs in this study, fits into this category. Then there are the so-called red-brick universities such as Manchester and Leeds, and the new universities of the 1960s like Warwick. These two strata within the HE sector are sometimes known as selector universities because of their popularity and status. The post-1992, or 'new', universities are frequently seen as teaching universities rather than research institutions and are sometimes specifically referred to as widening participation universities or recruiter universities. Bankside, the other HEI in this study, falls into this category. The HE system that has emerged in the twenty-first century is therefore heavily socially stratified. This stratification is widely understood and acknowledged both within the HE system itself and in wider discourses in the media. Elite universities receive better funding than the new universities, in response to their international reputation for research, and receive public acclaim in published international league tables. The new universities and the degrees they offer are routinely critiqued and have even been dismissed as offering Mickey Mouse courses by the media.

There has been massive change over the last 50 years in patterns of HE progression; the most dramatic of these have been the increase in numbers of undergraduates in the UK and that overall women now outnumber men. However, research suggests that the massification of HE in the UK and in other countries has not facilitated the most economically and socially disadvantaged young people in society to access HE. Indeed, the focus on encouraging under-represented groups to access HE has been a subject of critical discussion in many countries including the USA (Kahlenberg, 2004) and Australia (Chapman and Ryan, 2003; Gale and Tranter, 2011). Evidence suggests that in the mid- to late 1990s the gap in participation rates between the higher and lower social classes had widened in the UK (Blanden and Machin, 2004; Machin and Vignoles, 2004; HEFCE, 2005; Glennerster, 2001, in Vignoles and Crawford, 2010: 49).

The Labour governments' response to these issues was to develop a range of WP policies, targeting the HE sector itself through the introduction of the Office of Fair Access (OFFA) as well as young people in schools and colleges through a range of national initiatives, Aimhigher in England and similar schemes across the UK being the most expansive of these. These schemes, however, have been critiqued for their failure to widen participation among the lowest socio-economic groups (Galindo-Rueda *et al.*, 2004; Gorard *et al.*, 2007). According to the Higher Education Funding Council for England (HEFCE, 2010b), there was an increase of 5 per cent in participation rates since the mid-2000s from the most disadvantaged areas, and a reduction in differences between advantaged and disadvantaged neighbourhoods. However, entrenched differences remain, with fewer than one in five pupils from the most disadvantaged areas entering HE. Unsurprisingly, given the history of the HE sector, while there are more working-class and ethnic-minority students going to university than before, they are going to 'different universities to their middle-class counterparts' (Reay *et al.*, 2005: 9). As Leathwood (2004: 31) suggests, 'the hierarchy of universities both reflects and perpetuates social inequalities'. Minority-ethnic groups are now over-represented in HE, but there is a clear racial divide between the new and old sectors (Shiner and Modood, 2002; Modood, 2004; Connor *et al.*, 2004). The Sutton Trust (2007) outlines a polarization within the HE sector, with working-class and minority-ethnic students accessing different universities to their white, more middle-class counterparts. There are also variations between institutions in terms of meeting target numbers of students from under-represented groups. While the new universities 'generally performed at or slightly above their benchmarks' (Parry, 2010: 38), a number of other universities performed 'significantly worse than expected', suggesting that despite the raft of WP policies introduced by New Labour, the stratification within the HE sector remains entrenched.

The HE system that has emerged in the early twenty-first century is very different to the one many of the academics working within it experienced when they were students. Academics have been critical of the increased emphasis in HE on employability and vocational training and see this as at odds with the liberal-humanist tradition within HE of educating the whole person (Steele, 2000; Bridges, 2005). Janice Malcolm (2000: 20) suggests that the marketization of the HE system has contributed to students being isolated from traditional classed and gendered positions and has instead created 'individualized, self-directed, consumers of learning'. She comments that HE is increasingly 'a gathering of individuals out to get the best bargain for themselves' (ibid.); despite discourses of meritocracy, young people are clearly not equally positioned in this marketplace to do so.

Under the 2010 Coalition Government, this marketization of HE has progressed further. The White Paper (BIS, 2011) dramatically changed the financing of universities, with students becoming primary funders through paying increased fees. In this system, students are positioned as fully fledged consumers of HE who will 'take their custom to places offering good value for money' (32). With this white paper, the drive to increase numbers participating in HE in the UK ended, and WP policy was limited to a new National Scholarship Programme to support pupils from lower socio-economic backgrounds, while OFFA was set up to 'make sure institutions fulfil their outreach and retention obligations' (BIS, 2011: 7).

Widening participation and identity

A key tenet of Coalition policy in relation to HE is in the emphasis placed on students as consumers who can make informed choices. The Department for Business, Innovation and Skills (BIS) White Paper states emphatically that 'putting financial power in the hands of learners makes student choice meaningful' (BIS, 2011: 5). There are, however, well documented problems associated with positioning students as rational choosers of education who are equally positioned to compete in a marketized system (Hodkinson *et al.*, 1996; Ball *et al.*, 2000; Reay *et al.*, 2005; Archer *et al.*, 2003; Slack *et al.*, 2012).

The discourses of individualization, within which Coalition Government policy so clearly falls, disguise the extent to which young people are constrained by their circumstances. Beck (1992: 87) describes the 'social surge of individualization', as traditional social relationships have been replaced by more diversified lifestyles, with young people 'as the planning office with respect of his/her own biography' (135). However, the extent to which young people have any real control is highly questionable, as social structures continue to maintain and reproduce the social, ethnic and gender divisions so central to understandings of self and future expectations (Furlong and Cartmel,1997). Young people have an increased perception of their ability to make individual 'choices', despite the reality of their still being heavily constrained by social structures (Ball *et al.*, 2000). Ball *et al.* (2000: 4) suggest that, as a result of this, young people are likely 'to blame themselves for any lack of success'. Evans's (2007: 8) study of how young people experience control and personal agency during periods of transition also demonstrates that young people often ascribe failure to 'personal weaknesses' and to the problems of being 'only an average person in a highly competitive economic setting'. A criticism levelled against New Labour WP policies is that their aspiration-raising strategies also conformed to this discourse of individualization, with its emphasis on individual students and the need to raise their aspirations, rather than addressing the 'material poverty and social inequality' (Morley, 2003; Burke, 2012) that constrain them. Burke (2012: 105) describes aspiration-raising strategies as being 'largely constructed as an individual self-improvement project'. The resulting focus on the 'deficit' (Thomas, 2001) of non-traditional students has been widely criticized, because this focus problematizes certain groups of students and presents them as failing. This failure is counterposed with the gold standard of established middle-class student motivation and ambition (Burke, 2002; 2102; Archer and Yamashita, 2003; Yorke and Thomas, 2003; Watts and Bridges, 2004; Bridges, 2005). This focus on raising aspirations fails to pay due attention to complex 'processes of identity formation' (Burke, 2012: 101) that are constantly shaping and reshaping young people's subjectivities and ideas about possible future identities.

There has also been much criticism of the labelling of students as non-traditional (Burke, 2002; Burke 2006; Hockings *et al.*, 2010). Hockings *et al.* (2010: 195) highlight the importance of 'a multifaceted view of student diversity' which extends 'beyond the structural relations or divisions of class, gender and ethnicity'. In doing so, they draw on theories of intersectionality (Crenshaw, 1989; Mirza, 2008; Morley and Lussier, 2009). This is important if we are to move beyond the polarizing constructions of identity frequently found in media discourses relating to educational achievement. The current focus on white working-class male student underachievement often leads to unhelpful discourses of competition between the sexes and between different ethnic groups. White male failure is often juxtaposed with ethnic and female success. Indeed, David Willetts, as Minister for Universities and Science, suggested feminism and the education of women were responsible for 'holding back working men' (*Telegraph*, 1 April, 2011). The fact that women now outnumber men in HE in the UK has received much media attention, but subject areas continue to be highly gendered. Morley and Lussier (2009: 75) point out that women globally tend to be concentrated in 'subjects associated with low-wage sectors of the economy'. The whole picture is highly complex, and juxtaposing the failure of one essentially defined group with another (Ringrose, 2007) only obscures this.

The polarization and segregation of people according to their class and socio-economic status during formative periods contributes to differing levels of achievement between working-class and poorer people and their middle-class counterparts (Reay, 2001). The sense of difference this engenders then becomes entrenched in the minds of students; Reay *et al.* (2005) identify schooling as playing a key role in this by developing and consolidating this sense of difference. This acceptance of difference and experience of 'constraints framing the individual's earlier life experiences' can lead young people to 'accept exclusion or exclude oneself' (Reay *et al.*, 2005: 24).

Brooks (2003a: 283), however, urges caution over identifying middle-class students within one bracket and argues that understanding the 'decision making processes' of the 'lower middle classes' is important to developing a 'more nuanced account of how young people think about their futures'. It is interesting to note that the majority of parents of the sixth-form pupils in her study were not HE educated themselves, but that several pupils progressed to elite HE institutions. It is very clear from research evidence that young people from different backgrounds do not access information about university equally (Slack *et al.*, 2012).

Reay (2001: 33) describes the psychological cost to working-class students who do manage to progress to HE and suggests that the education system itself 'valorizes middle-class … cultural capital'; if a working-class student is to be successful in the education system, it is necessary for them to take on the attitudes and aspirations of the middle classes. In relation to students' experiences of university itself, it has also been highlighted that students whose parents have not themselves attended university have little knowledge at the start of their course about what is expected of them (Crozier *et al.*, 2010).

There is, then, considerable research pointing to the complexity of the structural and resulting psychological difficulties facing groups of young people moving through the education system. It is also important to engage critically with the whole focus underlying New Labour's WP strategy, which appears to assume that the aspirations of the established middle class are superior to those of their counterparts from lower socio-economic groups. Bridges (2005) argues that we should respect and value 'people for who and what they are'. It is important to interrogate assumptions that HE is 'the gold standard' for those with 'high levels of potential and ability' (Burke, 2012: 34). It may be that new apprenticeship schemes and higher apprenticeships in STEM industry may in time provide an alternative and equally desirable route for young people. The Coalition Government's focus on providing more, particularly online, information – 'wider availability and better use of information for potential students is fundamental to the new system' (Burke 2012: 32) – appears unlikely to positively impact on the decision-making of young people. In order to better support this decision-making process it is vital to understand what information is accessed and influential and how best to enable young people to access valuable and accurate information that will support them in making informed decisions.

Identity and learning

Critiques of the achievements of New Labour WP initiatives have often been based on statistical information and have focused on the lack of progress in changing the numbers of poorer and working-class students accessing university (Doyle and Griffin, 2012). Burke (2012: 70) notes 'an over-emphasis on collection of measurable data' in WP research and evaluation. It has been noted in educational research more generally that there is a 'new orthodoxy' of the authority of mainly quantitative research (Hodkinson, 2004: 10–1).

This focus on numbers also became central in practitioner research; given the critiques of WP initiatives, notably by Gorard *et al.* 2007, an understandable preoccupation developed with identifying quantitative and qualitative evidence of shifts in attitudes towards HE following attendance at events and activities. Hodkinson and Macleod (2007) suggest that a focus in research on the outcomes of learning such as test results, the 'static products of learning', is 'indicative of seeing learning as acquisition'. This critique is also applicable to the focus on outcomes in WP research. Viewing learning as acquisition is not the most useful way of conceptualizing learning in WP contexts; there is a need for a more 'nuanced understanding of teaching and learning in relation to questions about WP' (David *et al.*, 2010: 6).

Burke (2012: 70) identifies the need for methodologies in WP research that 'engage the subjective and experiential dimensions of cultural and subjective experiences, identities, relations and inequalities in different educational and pedagogical contexts'. Another critique of widening participation is that it has been 'too focused on classroom activity' and that 'more informal learning such as social learning has not been sufficiently investigated' (Stuart, 2006: 181). Indeed, as only 18 per cent of most children's time awake is spent in school, contemporary society's preoccupation with school as the 'sole site of learning' (Osborne and Dillon, 2007: 1441) appears to be somewhat misplaced. My focus is the learning that generally takes place outside the classroom in the various contexts involving groups of pupils and student ambassadors during outreach activity; understanding the subjective experiences of young people and social processes in the learning that is taking place is crucial.

There have been various studies about how young people make choices, and much of the literature points to the importance of informal relationships in providing young people with information. 'Hot' knowledge is described as the 'grapevine knowledge' that people access informally through their social networks. This knowledge is described by Ball and Vincent as 'immediate', and appears more important to those in their study than 'the "cold" formal knowledge' produced by institutions. 'Grapevine knowledge' is 'socially embedded'; the socially embedded nature of these sources is at the root of their being viewed as more trustworthy and reliable than other more removed sources. Various studies (Ball and Vincent, 1998; Ball *et al.*, 2000; Reay *et al.*, 2005; Archer *et al.*, 2003) have identified that working-class students and their parents rely more heavily on 'hot' sources of information, grapevine knowledge, than their middle-class counterparts who more easily access official or formal sources.

Friendship groups are potential sources of 'hot' knowledge about HE. Reay *et al.* (2005: 154) provide examples of how particular peers or individuals can influence students' choices; the mature students in their study 'seemed to be swayed by personal recommendation unsupported by other evidence'. Stuart (2006) explores how friendship groups are explicitly responsible for providing the information and support necessary for students from lower socio-economic groups to decide to progress to HE in the first instance. This reliance on 'hot sources', though, is seen to limit working-class students' horizons of action. Stuart (2006: 175) describes how vital friendship groups are for the mature students in her study in providing 'support and knowledge', as there are 'few academic supports in their lives'. She suggests that 'teachers and tutors do not seem to fill this gap' and so it is students' friendships that 'powerfully affect these students' lives' (ibid.).

However, Brooks (2003a; 2003b), reveals that for the groups of lower-middle-class sixth-form pupils in her study, friendship groups are not comfortable places for discussions about HE decision-making. She explores how such conversations tended to be avoided as they highlighted 'significant differences between friends and others in the wider peer group' (2003b: 237). The friendship groups of pupils and wider peer group, though, operated powerfully on pupils' HE choices in other ways. The academic hierarchies within which pupils viewed themselves served to define their selection of type of HE institution and, in some cases, choice of subject. Also important are pupils' self-perceptions as 'high achievers', developed early on during the course of their education (Brooks, 2003a). Reay *et al.* (2009) also point to the importance of established successful learning identities to the academic success of students at university.

According to Reay *et al.* (2005: 152), it is not just the 'working classes' who are influenced by 'hot' sources of information; the most affluent and privileged students and their parents also give 'primacy to hot knowledge', they are just privy to more elite 'hot' sources than their more working-class counterparts. The lower-middle-class pupils in Brooks's (2003a) study were generally the first generation to access HE, but Brooks identifies the 'specific context within which their parents worked' as contributing to their knowledge about HE and HE hierarchies, with some parents having access to more detailed knowledge about how different HEIs privilege their graduates. Brooks stresses the correspondence between pupils' and their parents' views of HE and that pupils and their parents were 'actively attempting to change their habitus' (295).

Evidence also indicates that visits to universities and meeting students and teachers have an important impact on young people's decisions to apply to HE (Slack *et al.*, 2012); it also highlights how their perception of ambassadors as 'people like me' (Archer *et al.*, 2003) are also influential. Slack *et al.* (2012) identify the importance of 'warm' sources of information, 'previously unknown university students' met during university open days, to young people's decision-making about university.

In response to these insights about the importance of such informally obtained 'hot' and 'warm' knowledge' about HE and about the role of friends, family and peers in informing pupils and in locating them as belonging to certain stratifications within the HE sector, I have been motivated to explore ideas about learning as a social activity. Several years ago, in a paper co-authored with Anna Paczuska, we considered the role of learning within the ambassador relationship in relation to a young person aspiring to become a member of a community (Gartland and Paczuska, 2007). We suggested that one way to study what motivates human beings to engage in interactive and social activity so they can learn is to look at how they see the learning they engage in as being useful to them in terms of social outcomes, and not just in terms of academic achievement per se. Learning is a pre-requisite stepping stone for gaining something which is important to them in its own right. One area in which gaining academic skills may be of use is in fulfilling a desire to become members of a privileged group. We suggested that one privileged group to which young people may wish to gain access is that which their student ambassadors or mentors represent, namely university students. The work of Lave and Wenger (1991) on apprenticeship models of learning was relevant to this analysis. They argue that learning is 'situated' and is a function of the activity, context and culture in which it occurs. They identify social interaction as a critical component of situated learning. Certain learning, such as specific beliefs and ways of behaving, serve to give learners recognition in a 'community of practice'. Newcomers move from the edges of a given community to becoming an accepted and engaged member of the community as they gain knowledge of that community, and act out its values and processes. They move from apprenticeship to being an expert member of the community, acquiring the knowledge necessary to become a member of the 'community of practice' largely informally and unintentionally through 'situated learning'. The expertise gained is key to gaining membership of the group and for social acceptance as part of that group. While the learning itself is valuable, its primary importance is in the status it confers on the learner in enabling them to gain admission to a social group.

This way of thinking about learning appeared apposite to the learning that occurred within student ambassadors' relationships with school and college pupils. If learning for academic success is viewed not simply in terms of the skills it confers, thus enabling a young person to progress academically, but also in terms of enabling them to gain their first foothold in a new community to which they seek entry, the 'community of practice' of university students, then the dynamics of the social process which motivates that learning, rather than the learning itself, gains significance.

However, there are also other useful ways of thinking about the learning processes that take place in these contexts. Hager (2005) is critical of conceiving of learning as participation, arguing that this fails to emphasize that learning is a process, not something that is static. He contends that learning is most effectively viewed as a process of construction. Evans *et al.* (2006: 13) provide an extended theory of situated learning to consider current practices in workplace learning, which include recognition of individual biographies as being significant to 'engagement in learning environments'. As

Hodkinson and Macleod (2007: 6) suggest, the view of learning as 'embodied construction' and 'as taking place through participation in learning cultures' are 'mutually complementary' and 'can and should be combined'. Hager and Hodkinson (2011) propose that a useful way to think about the learning process is to think about 'learning as becoming', a metaphor that combines these theories of learning.

This metaphor presents learning as being a process of developing identity rather than simply about acquiring information for particular purposes. If we consider this conception of learning as 'becoming', it is useful to consider post-structuralist thinking about subjectivity. Post-structuralists move away from the liberal-humanist conception of people as being autonomous individuals to an understanding of a subjectivity that is created through available discourses and ways of being embedded within particular cultures and moments in history. Judith Butler's ideas about 'performativity' are of particular relevance here. Her work suggests that identity is performed and that we become who we are through the ways we speak and behave – that our identity is 'constituted through action' (David *et al.*, 2006). As Burke (2012: 109) suggests, 'aspirations are relational ... they are formed in relation to others'; Butler highlights the importance of social acts in the process of becoming a subject:

> At the most intimate levels, we are social; we are comported towards a 'you'; we are outside ourselves, constituted in cultural norms that precede and exceed us, given over to a set of cultural norms and a field of power that condition us fundamentally.
>
> Butler, 2004: 45

Drawing on Butler's theories, Davies (2006), in her analysis of the ways in which primary school teachers position and 'constitute' their students, suggests that the power wielded by those in positions of authority is unacknowledged because of the emphasis in educational thinking on the learning process of the individual:

> The responsibility and power to shape students inside the range of possible subjectivities, subjectivities that are recognized as viable ways of being, are thus papered over in this emphasis on the freedom of the subject who is actively shaping itself through engagement with the syllabus ... .
>
> Davies, 2006: 430

If we consider Butler's theories and Davies's analysis we are again led away from separating the social learning that occurs among peer groups from the learning that occurs in more formal contexts. Davies considers Butler's theory of 'subjection' within the context of primary education and provides accounts of how pupils attempt to subvert power relations imposed by their teacher by taking up positions that provide successful male identities within their peer groups. Such identities, however, are not acceptable within the school context:

> Sometimes we are caught between more than one meaning system. The naughty boys successfully displayed dominant heterosexual masculinity in the playground, only to find themselves humiliated and shamed, and unrecognizable as viable in the teacher's eyes unless they gave up such accomplishing of dominant male identity.
>
> Davies, 2006: 435

It is easy to see how learner identities are constituted early in a child's educational journey, and that pupils with successful learner identities, like those in Brooks's and Reay *et al.*'s studies (respectively 2003a, 2003b and 2004; 2005 and 2009), progress and succeed, while those who do not establish these learner identities early on are likely to struggle throughout. Issues of identity are, then, of central importance when considering learning and teaching and HE access. In order to understand how best to support young people in progressing into and through HE, we desperately need to understand the myriad of influences constituting and refining these young peoples' identities.

Chapter 2

Widening participation and STEM: the cases of engineering and medicine

The motivation behind much outreach and WP work in STEM is about ensuring there are suitably qualified undergraduate applicants in subject areas that are seen as vital to the economy. The entrenched gender division in some STEM subjects and jobs remains a substantive issue, and debates surrounding pupil engagement with STEM, as well as questions about the gender divide and the representation of different ethnic groups are a continuing issue for work directed at engaging students with these subjects.

STEM and the economy

The importance of STEM subjects to the UK and international economies has been highlighted (National Research Council, 2010; Little and Leon de la Barra, 2009; Leitch, 2006; Lambert, 2003; Sainsbury, 2007; UKCES, 2009; CBI, 2010; Perkins, 2013). Although the term STEM is sometimes inclusive of medicine, the focus of concern is often on science, technology, engineering and maths subjects. In the UK, the CBI, in their 2010 education and skills survey, outline the urgent need for these STEM skills by businesses, with 72 per cent of companies saying that they employ STEM-skilled staff. The report identifies STEM skills as 'vital to areas of future growth and employment' and observes that 'skills shortages may hold back progress', with '45 per cent of employers currently having difficulty recruiting STEM skilled staff' (CBI, 2010: 7). The UK Commission for Employment and Skills (UKCES) predicted that 58 per cent of all new jobs in the UK will be STEM-related so there will be 'significant growth in jobs but also massive replacement demand' (UKCES, 2009: 92). They suggest that 'economically valuable skills (intermediate and higher STEM skills) will matter most'. A report from the Social Market Foundation pointed to a likely shortfall of 40,000 home graduates for STEM-sector jobs (Broughton, 2013).

The primary driver for widening participation in engineering and other STEM subjects, in the UK and internationally, has been a perceived need to find new talent to sustain economies. Connected to this is the 'instrumental value' of STEM knowledge due to an 'increasing dependence of contemporary society on sophisticated artefacts' making us 'communally dependent on individuals with a high level of scientific and technological expertise and competence' (Osborne et al., 2003: 1052).

There are other motivations. A range of benefits to having a more diverse community of engineers has been identified. In 2009, the President of the Royal Academy of Engineering (RAEng) outlined the importance of widening participation in engineering, suggesting that 'diverse teams produce better results in engineering', as 'different experiences and ways of thinking often lead to innovative outcomes'. He also observed that the profession should 'reflect the diversity in our society', as this will enable engineers to better 'understand, communicate and engage effectively with the wider community' (RAEng, 2009: 1). Professor Matthew Harrison (2011), Director of Education at the RAEng, identifies the contribution of STEM careers to social mobility. Research undertaken by the Institute of Education (Greenwood et al., 2011) indicates that there can be a wage premium associated with STEM qualifications, particularly in technology and engineering, supporting the claim that currently under-represented groups could benefit from gaining STEM qualifications.

In contrast to engineering, there is no concern about recruitment of young people to careers in medicine. Undergraduate courses in medicine are regularly oversubscribed. Indeed, the HEFCE advisory group report on strategically important and vulnerable subjects (SIVS) suggests that the growth in popularity of disciplines like medicine and related subjects is likely to have contributed to the comparatively lower levels of recruitment in other STEM disciplines including engineering (HEFCE, 2009b: 17).

Recruitment to medicine and engineering subjects
STEM subjects differ significantly in their patterns of recruitment to university courses.

Medicine
Medicine is widely considered a prestigious subject to study at university (Brooks, 2003a) and courses are continually oversubscribed. Women now outnumber men on undergraduate courses (Boursicot and Roberts, 2009). This is not to say that gender is no longer an issue in medicine, as the higher levels of the profession are still male dominated; it does, though, represent a real shift over the last three decades.

However, progress in terms of representation of different minority-ethnic groups within medicine has been less clear in the UK. Minority-ethnic students make up 35 per cent of undergraduates in medicine and dentistry (Connor *et al.*, 2004). Students from Asian backgrounds are over-represented in proportion to their percentage of the population as a whole; between 10–14 per cent of all candidates accepted into the medical sciences in 2009 were of Indian origin (Smith and White, 2011), but other groups have not fared so well. Black African and Caribbean students are significantly under-represented on medical degrees (Bouriscot and Roberts, 2009; Greenhalgh *et al.*, 2004; Connor *et al.*, 2004). Male Black African students only make up 0.7 per cent of undergraduates in medicine and dentistry; students from lower socio-economic groups are also still significantly under-represented (Grant *et al.*, 2002). In 2008/9, only 4.5 per cent of entrants to full-time degree courses in medicine, dentistry and veterinary sciences were from low-participation neighbourhoods, according to the Higher Education Statistics Agency (HESA, 2008/09), and a high 60 per cent of students studying medical sciences in 2008 were from higher and lower managerial and professional occupational classes (Smith and White, 2011).

The Council of Heads of Medical Schools (CHMS) issued a policy statement in 2004 identifying the need for graduates to reflect the diversity of the patient population. However, as Boursicot and Roberts (2009: 22) suggest, though well meaning, such 'statements do not necessarily ensure that changes are implemented to make the medical profession more inclusive or less elitist'.

Engineering

While medicine is oversubscribed, engineering has been identified as a strategically important and vulnerable subject (SIVS). Findings of the HEFCE advisory group about SIVS (HEFCE, 2009b: 3) suggest that, even though numbers of students in chemistry, physics and mathematics are increasing, albeit at far lower rates than medicine, numbers of students in engineering programmes continue to decline as a whole; there are, however, differences between sub-disciplines, with civil and chemical engineering numbers actually increasing.

Besides the differences in terms of levels of recruitment, there are also differences between medicine and engineering in terms of the gender of students who apply. Unlike medicine, undergraduate engineering courses in the UK still attract predominantly male students. There is quite a high proportion of middle-class students, though not as high as in medicine; in 2008 39 per cent of students were from higher and lower managerial occupational classes (Smith and White, 2011). In 2008/9, some 8.2 per cent of entrants to full-time first degree courses in engineering and technology subjects were from low-participation neighbourhoods (HESA, 2008/09). In terms of minority-ethnic participation more generally, engineering courses recruit better than some other related disciplines, but not as well as medicine. Just over 20 per cent of undergraduates in engineering and technical subjects were drawn from minority-ethnic groups, compared to only 10 per cent in the physical sciences. However, representation of minority-ethnic groups is uneven, with African and Asian male students better represented than other groups, particularly in electronics engineering (Connor et al., 2004; HESA, 2009/10). Asian students represented 10 per cent of engineering students accepted in 2008 (Smith and White, 2011).

Though there has been some improvement in the percentage of women studying engineering in the UK, the numbers are still very low. Twenty years ago, 4 per cent of the undergraduate population was female; this figure was at 13 per cent in 2009 (RAEng, 2009). Female students are severely under-represented in all engineering subjects, though this is more pronounced in some subject areas than others. They constitute only 7 per cent of students studying mechanical engineering, but 22 per cent of chemical and process engineering (Engineering UK, 2011). It is interesting to note that 18 per cent of women studying chemical and process engineering in 2008/9 were Black African, which may be linked to the presence of petrochemical companies in Nigeria (HESA, 2008/09).

This lack of participation is not unique to the UK. Engineering remains a male-dominated profession across the world (Morley, 2012). Women constitute between 17 and 20 per cent of engineers in the US, China, South Africa, Sweden and Portugal, and in Switzerland, Germany and Japan just 2 per cent or less (Baker et al., 2007). In the US, women as well as African Americans, Hispanics, Native Americans and some Asian American groups are significantly under-represented in engineering (NAE, 2008).

Connor et al. (2004) note the higher proportion of Black African students compared to other groups choosing subjects with particular career or job outcomes in mind. It may be that the parents of students of African heritage have more positive perceptions of engineering than some other minority-ethnic groups and white pupils. The influence of parents generally on HE decision-making (Brooks, 2004) and on the decision-making of minority-ethnic groups in particular (Connor et al., 2004), as well as on the formation of science identities (Archer et al., 2012), has been noted.

In the UK and globally, there has been a move in engineering HE education towards more active project- and problem-based learning (Sainsbury, 2007; Northwood et al., 2003; Arlett et al., 2010). This has been driven by the changes in the skill sets required of engineering graduates to meet the needs of industry (Sainsbury, 2007; RAEng, 2007) and the related need to recruit and retain more diverse students into engineering careers (Arlett *et al.*, 2010). Engineering is following in the footsteps of medicine subjects where problem-based learning (PBL) is well established (Williams and Lau, 2004; Northwood *et al.*, 2003).

A problem identified for engineering in particular has been that there is little understanding among the public generally about engineering and what engineers do. In a survey of perceptions of engineers (Engineering UK, 2011: 48) 20–40 per cent of 20-plus year olds were said to know 'nothing' about 'specific types of engineering, or of the day-to-day realities of what various roles involve'. The report identified that 28 per cent of 20 year olds agreed that 'hardly anyone knows what engineers do'. The survey also found that there was confusion around educational pathways into the profession and that engineers were perceived to earn less than other professionals, such as those working in medicine, accountancy or law. School pupils often describe engineers as being people – and particularly men – who fix things, such as car mechanics, computer engineers or electricians (Canavan *et al.*, 2002). In 2002, the Engineering Education Alliance (EEA) of 30 engineering organizations, including the RAEng and SEMTA (the skills council for science, engineering and manufacturing), was formed in order to 'provide a coordinated, simplified and consistent approach to educational initiatives related to engineering' (EEA, 2002: 1). One key focus for this alliance was 'simplifying and promulgating the "big-picture" engineering message' (ibid.). Its report suggests that, in the UK, there are too many different messages about engineering coming from different bodies, and that these can seem conflicting and confusing and do not include 'all sectors of society' (4).

In the US there are similar issues, and considerable amounts of money have been spent on challenging public perceptions. The National Academy of Engineering (NAE) revealed that the 177 organizations involved in developing public understanding of engineering spend an estimated 400 million dollars annually (NAE, 2002). Despite this investment, there appears still to be limited understanding of the role of engineers (Cunningham and Knight, 2004; Cunningham *et al.*, 2005; 2006). Research suggests that the US public see engineers as being less engaged with wider social issues than other professions, and that engineering is viewed as less prestigious than professions such as medicine, nursing and teaching (NAE, 2008).

The history of routes into medicine and engineering

Boursicot and Roberts (2009) describe a 'strong conservatism' within the UK medical profession dating back to the nineteenth century, when medieval guilds controlled who entered the profession. In their view, this system has maintained 'pre-19th century hierarchical distinctions between practitioners and educational establishments' and has perpetuated the traditional elitist position of the medical profession (Boursicot and Roberts, 2009: 20). This position is seen as being entrenched by the 1858 Medical Act that saw the creation of the General Council for Medical Education and Registration, later known as the General Medical Council (GMC). Members of the council were appointed by the Crown, universities and medical corporations. This had the effect of 'consolidating the powers of protectionism of the different sub-groups of the medical profession'. While universities delivering medical degrees are subject to higher education legislation and policy, 'the content of the courses, the learning objectives and outcomes ... are set out in guidance documents produced by the GMC' (Boursicot and Roberts, 2009: 21). Boursicot and Roberts argue that today these systems impact on the practices of institutions and contribute to the lack of equity in admissions. They cite Allen (1988); McManus (2002); Bowler (2004) and Goldacre *et al.* (2004), and suggest that 'institutional racism ... still seem[s] to pervade the medical profession and the National Health Service' (ibid.).

The history of engineering within universities is neither so long nor so elitist. The Fellowship of Engineering – later to become the Royal Academy of Engineering in 1992 – was only established in 1976. Chemical and mechanical engineering flourished as academic disciplines in universities during the nineteenth century due to the Industrial Revolution. Electronics also emerged as an academic discipline during this time. However, the subject's association with industry ensured that engineering never attained the prestige of traditional disciplines like medicine, though it became well established at elite institutions. There were always other routes into engineering careers, and apprenticeships provided access to engineering to cohorts of society other than the upper middle classes. Technical subjects were taught in polytechnics and technical colleges set up in the nineteenth century. These colleges later became polytechnics in the 1960s and delivered many engineering degree courses. These different routes provided and still provide opportunities for more diverse groups of students than were available in medicine. Indeed, in large companies there are still routes into senior positions for those who enter even as apprentices. This is reflected in the fact that some senior managers of companies such as BT and Tube Lines progressed through the company in this way.

Given these differences in terms of the history and structural obstacles for entry into the two professions, it is, perhaps, surprising that engineering has failed to appeal to more diverse groups of students. However, as is the case with medicine, the contents of engineering degrees are constrained. Pre-1992, the Council for Academic Awards (CNAA) used to accredit the degrees to the former polytechnics, but post-1992 the universities accredited their own. This meant that the value of the award became closely tied to the status of the university itself. Because of this, the importance of the accreditation of professional bodies has increased. This is seen to have been problematic for new universities in particular, as some professional bodies hold a traditional perspective about the appropriate content of engineering courses and are seen as constraining new developments.

STEM identities, school science and the gender divide

Medical degrees are extremely competitive and high grades are demanded of candidates applying to courses. Brooks' (2003a) findings about subject hierarchies in young people's higher education choices reveal that the sixth-form pupils in her study viewed medicine and law as subjects for high-achieving pupils. She suggests this is linked to the status of careers in these subjects. Indeed, the majority of pupils in her study had 'been keen to study medicine' at the start of their sixth-form education (Brooks, 2003b: 246). In contrast, applied or vocational subjects were valued by one pupil below humanities courses, which he thought provided a 'better all round education' (Brooks, 2003a: 286). Others valued courses for their links to a 'particular employment sector' (ibid.). Brooks comments that there is little research about 'the ways in which hierarchies of degree subjects are constructed' by pupils or how these hierarchies 'intersect with perceptions of institutional status' (ibid.). There is an issue for engineering and other science-related subjects, which do not have the same career status in young people's eyes as medicine, and so are not aspirational subjects for high-achieving pupils.

While HE courses in physical sciences and engineering are comparatively under-subscribed, applicants for any of these subjects still need STEM qualifications at Key Stage 5 (KS5) of the UK National Curriculum, and to access the elite institutions students still need to attain high grades. Any attempt to widen participation in engineering or medicine is, then, clearly circumscribed by the high drop-out rate from school science after Key Stage 4 (KS4) and the comparatively low numbers taking up STEM subjects at KS5. As Osborne and Dillon (2007: 1442) observe, the impact of learning in informal contexts is likely to be substantial; however, research into this remains in 'its infancy'.

The problem of lack of student engagement with science has been ongoing for three decades (Ormerod and Duckworth, 1975; Schibeci, 1984; Osborne *et al.*, 2003). Lack of engagement with mathematics at KS5 is also problematic for recruitment to HE, particularly in the physical sciences and engineering. Williams *et al.* (2010) identify schools' focus on league tables and results as a key issue promoting an 'exchange value' view of mathematics, which in turn promotes identities among pupils as 'surface learners' of the subject. Another issue for STEM subjects is the school curriculum, which presents a 'backward-looking view of the well-established scientific landscape', when what 'excites students is the "white heat" of a technological future offered by science' (Osborne *et al.*, 2003: 1062).

These issues affect both genders, though a particular concern has been the lack of gender parity in engagement with STEM subjects. A recent survey conducted by researchers at the Institute of Education (Reiss *et al.*, 2010) illustrates the gender divide in students' intentions to continue with maths and physics post-16. The survey of 10,355 14–15 year olds at 113 secondary schools found there were significant differences in students' intentions to pursue physics and maths at A level. These students shared similar background characteristics, but their orientation to physics and maths revealed a sharp gender divide. Only 5 per cent of girls expressed the desire to pursue physics compared to 13 per cent of boys. In maths, the gender divide was less pronounced but still very evident, with 22 per cent of boys expressing an intention to pursue the subject, compared to 15 per cent of girls.

This gender divide in STEM subjects appears entrenched despite the work of a range of national organizations to support women's progression in science. These include WiC (Women in Computing), the Women into Science and Engineering (WISE) campaign, the UK Resource Centre (UKRC) for women in science, engineering and technology (SET) and the ATHENA Programme for women in SET in HE. Funding cuts are also impacting on the work of these organizations (the UKRC has lost all government funding).

Wide-ranging views have been expressed about the reasons behind this gender imbalance. Francis (2002: 77) argues that children's choices reflect 'a gendered dichotomy' and that boys are more likely to choose jobs 'involving technical, scientific and business skills', with girls choosing jobs that involve 'creative or caring elements'. Eisenhart *et al.* (1996) argue the stereotypes held of scientists as 'nerdy, male and white' act as a deterrent to girls. Various studies have focused on the masculine status of scientific knowledge (Harding, 1991; Kelly, 1987; Watts and Bentley, 1993; Hughes, 2000).

Siann and Callaghan (2001: 91) outline a number of 'barriers' that have been explored in recent years: 'the masculine culture of SET', 'the image of the Scientist' and 'the lack of role models and female networks in SET'. However, Siann and Callaghan are critical of this focus on 'barriers' to careers in SET, commenting that 'programmes based on such critiques have not succeeded in women's participation in SET' and pointing out that 'a number of barriers cited have been characteristic of other occupational areas' (Siann and Callaghan, 2001: 91); they point to medicine as an example of where women have overcome such obstacles. Their argument is that female students are actively not choosing STEM, and they provide various reasons why this might be the case. They identify female students as more concerned with people and 'the needs of society', which may put them off technical jobs, and suggest that the pay, employment status and image of jobs may not appeal to women (Siann and Callaghan, 2001: 92). This reflects, to some extent, the findings of Lightbody and Durndell (1996), who argue that women are not interested in technical subjects as they are more motivated by subjects like medicine and law. However, the fact that school science and technology subjects do not effectively relate to exciting real-world applications may be contributing to this lack of interest (Osborne *et al.*, 2003; Watermeyer, 2012).

Various studies reveal how the workplace reinforces and embeds gender dichotomies (Rainbird, 2000; Evans, 2006). In her analysis of gender and vocational education, Evans (2006: 6) considers how gendered patterns of participation in different areas of work are reinforced and reconstructed. She suggests that these patterns are linked to social expectations and social 'normalcy'. Employees are then trained in 'gender typical competences' and this then leads to reinforcing and 'reconstructing' gender segregation in work places as 'occupational tasks and cultures influence ways in which "key competences" are recognized and deployed'. The findings of Beck *et al.* (2006: 682) suggest that school pupils conform to 'stereotypical notions of what men and women should do'. Watermeyer (2012: 679) describes how gender inequality in science is 'normalized within the institutional structures of formal education and in the inherited prejudices of educators'. Indeed, research suggests that those women who do take up engineering are influenced to do so by family, often having fathers who are engineers (Newton, 1987; Evans, 2006). Various studies (Francis, 2002; Beck *et al.*, 2006; Evans, 2006) indicate that female pupils' lack of access to engineering and technical careers disadvantages them in the labour market, a view supported by indications of labour market advantage experienced by many with STEM qualifications, particularly in technology and engineering (Greenwood *et al.*, 2011).

Hughes (2001: 276) suggests that focusing on gender is in itself problematic, as it disguises the complexities of 'other power relations based on ethnicity or class'. She argues that drawing on post-structuralism allows for a more nuanced analysis, as this approach recognizes that students 'actively negotiate subject positions within discursive constraints'. She suggests that students may 'take up science identities' or may reject them and that this depends on available discourses and practices within science itself:

> If being a scientist is congruent with gender subjectivities available within dominant discourses and practices of science, a scientist identity is relatively easy for students to construct. However, if the subjectivities are not compatible, a science identity is uncomfortable and may be rejected.
>
> Hughes, 2001: 278

Hughes (2001: 279) observes that other subjectivities relating to discourses of 'class, academic achievement and ethnicity will interact with scientist subjectivities', and so the factors affecting the possibility of students taking up or rejecting science are dependent on complex interactions between different aspects of students' identities.

Archer *et al.* (2010) again draw on post-structuralism in their analysis of primary school children's orientation to science. They observe that there has been very little work focusing on the 'views young students hold' about science and, like Hughes, suggest that there is a need for more qualitative work that 'understands learning as tied to processes of identity construction' (Archer *et al.*, 2010: 2). They identify that 28 per cent of students in their study start forming career aspirations as early as primary school and 35 per cent between the ages of 12 and 14, and argue that there is a need to consider the formative influences on younger students' aspirations. In their analysis of children's constructions of science, they make a useful distinction between 'doing science' and 'being a scientist' and identify that, while many children enjoy 'doing science', for many 'being a scientist' is 'unthinkable' (Archer *et al.*, 2010: 5). They suggest that the level of interest and engagement children have with science is 'shaped by their social structural locations', and that within these locations there are 'specifically class and racialized masculine/feminine identities' that young people 'see as desirable and constitutive of the self' (Archer *et al.*, 2010: 3):

> Its [science's] boffin associations and incongruence with popular/desirable forms of contemporary masculinity and femininity (especially working-class configurations) make it a potentially risky identity being closely associated with markers of an 'uncool' identity.
>
> Archer *et al.*, 2010: 18

In considering what young people 'see as desirable and constitutive of the self', it is necessary to develop an understanding of the 'viable ways of being' available to young people (Davies, 2006: 430). Butler's analysis of gendered identity and conception of a 'matrix' of 'cultural intelligibility' is also useful:

> I use the term heterosexual matrix ... to designate that grid of cultural intelligibility through which bodies, genders and desires are naturalized ... a hegemonic discursive/epistemological model of gender intelligibility that assumes that for bodies to cohere and make sense there must be a stable sex expressed through a stable gender.
>
> Butler, 1990: 151

Hey's (1997) study of girls' friendships reveals how friendship groups exert, in Foucauldian terms, a 'disciplinary' force over their peers. In her account, girls' 'occupation of and insertion into cross-cutting multiple regimes of power' leads to the construction of 'identifications against, as opposed to with, other girls' (Hey, 1997: 131). According to Walker (2001: 81), girls studying engineering at university 'construct themselves as in some way "different" from other girls'. In this construction of their difference, however, they appear to take up more normative masculine identities in a 'licensed mimicry' (McRobbie, 2006: 10) of their male counterparts.

If taking up STEM identities comes at the cost of giving up normative feminine identities, it is perhaps unsurprising that the take up of these identities is confined to small numbers of middle-class girls, 'one class defined constituency', who have the 'capacity ... to suppress aspects of (their) femininity and sexuality (Watermeyer, 2012: 682). Like Renold and Ringrose (2008), I am interested in the complexities of how girls reinscribe and disrupt this normative gendered identity within STEM.

The need to widen participation in STEM is driven by economic concerns about the future prosperity of the UK and has been a high-profile and long-term agenda for government. The process of engaging more diverse groups with STEM subjects also has been a long-standing challenge. There are clearly structural problems that work to exclude students, including course content and selection processes. Pupils' perceptions of which degree courses are valuable are also complex, and pupils and their parents do not value careers in science, engineering and technology as they do medicine. It seems that orientation to STEM subjects is intimately connected to the identities of young people, and that contemporary discourses powerfully define and constrain gendered, classed and ethnic identities, so that many young people simply do not consider a future relating to science, technology, engineering or maths as viable or even thinkable.

While there appears to be a comfortable fit between medicine and traditional female identities as carers, this is not the case for other technical and scientific jobs. For engineering in particular, the lack of accurate public knowledge and limited knowledge of engineering and engineering roles in society is particularly problematic. A way forward for widening participation in science is to 'disrupt dominant discourses around science and the identity of the scientist' and 'interrupt dominant identity patterns of (dis)identification' (Archer *et al.*, 2010: 21). The question of how this is to be achieved is yet to be resolved.

Chapter 3

Student ambassadors and mentoring

At the start of this century there was an explosion of student mentoring and ambassador schemes within UK universities, reflecting a national and international focus on mentoring as part of government strategies to address social and educational issues (Colley, 2003). In 2002, the Universities for the North East (UNE) mentoring projects directory listed 140 mentoring projects nationally, in which university students provided support to younger students and potential university applicants in various forms.

Undergraduate and post-graduate student employees were increasingly used by universities in WP outreach work with school pupils under successive Labour administrations. While there is no very clear definition of what a student ambassador is, due to the range of titles given to student employees working in universities and the variation in their activities (Sanders and Higham, 2012), 'student ambassadors' and 'student mentors' are the most commonly used terms. One other widely used term is 'student associate', which has generally been used in relation to particular schemes, especially the national Aimhigher Associates programme. A distinction is often made between mentors and ambassadors in terms of the apparent intensity of the relationship between them and the school pupils they work with. Mentoring is generally used to describe activities within one-to-one relationships between a young person and a university student. Activities undertaken in mentoring relationships may cover coaching or tutoring in a given subject or skill area to increase attainment, networking to increase a mentee's social capital, and activities designed to boost confidence and self-esteem. Where ambassador work is referred to separately, it usually implies a less intense relationship between a university student or group of students and school or college pupils for activities such as educational visits, summer schools or subject-based project work.

The distinction between ambassador work and mentoring is therefore not total, rather there is a general continuum of activities between one-to-one activity on the one hand, through group work on projects to group work with young people at one-off educational events. Ambassadors often work as mentors and mentors as ambassadors, and both mentors and ambassadors worked as part of the Aimhigher Associates programme. 'Student ambassador', though, is 'frequently used as a catch-all term to describe HE students working on outreach activities' (Sanders and Higham, 2012: 12).

Whatever their title, ambassadors, mentors and differently labelled student employees are widely viewed as efficacious in WP activity. While there is some lack of clarity about the exact role and purpose of student ambassadors, they are ubiquitously held to provide school pupils with role models and to be effective in raising aspirations in relation to HE. The majority of ambassador activity is with pupils in Years 9, 10 and 12 (Sanders and Higham, 2012) and there is a growing acceptance of the fact that young people need to be involved in 'aspiration activities' early on in their education (Kerrigan and Church, 2011: 36).

As well as working in WP outreach with school pupils, ambassadors also play a part in marketing HEIs and particular courses; they act as tour guides, work at reception dealing with enquiries and provide information about different courses. Ambassadors' contribution to working with students transitioning into HE and to the retention and success of undergraduates from diverse backgrounds has also been highlighted (Sanders and Higham, 2012; Andrews and Clark, 2011; Andrews *et al.*, 2012; Thomas, 2012).

Widening participation and ambassador work: evidence of benefits?

The individualized aspiration-raising discourse dominant in New Labour WP policy was clearly present in policy and practice relating to student ambassadors. There appears to be a consensus among those working in WP and government bodies that student ambassador schemes and other similar schemes were successful in supporting the Labour governments' WP agenda. The HEFCE 2005 *Evaluation of Aimhigher: Excellence Challenge Interim Report* draws heavily on this aspiration-raising discourse, identifying that:

> One to one contact with undergraduates either through a mentoring or other programme, emerged as a significant factor associated with higher levels of attainment and higher levels of aspiration in both the statistical and the qualitative studies
>
> HEFCE, 2005: vii

This report explains that coordinators commend 'the way in which interaction with higher education students' can 'play a part in breaking down cultural barriers' and the way in which ambassadors can make higher education 'cool in the schools' (HEFCE, 2005: 38). An evaluation of the Student Associates scheme for the Training and Development Agency (TDA) is also cited in this report: 'by being close in age and experience, Student Associates can relate to the issues young people face' (HEFCE, 2005: 38).

Indeed, the expansion of the new Aimhigher Associates scheme in 2009, after a pilot phase, to cover the whole of England reflected New Labour's belief in the value of this contact. Within this scheme, Student Associates, who were theoretically at least from state schools themselves, were supposed to work with small groups of specifically targeted pupils, according to their socio-economic status or other specific criteria, from Key Stages 4 and 5 over an extended time frame. The focus of activities was to 'help learners from the lowest socio-economic categories to progress to the full range of higher education provision on offer', *Handbook for Aimhigher Associates* (HEA, 2009–10), both by raising aspirations and by providing practical support. The foreword to HEFCE's *Aimhigher Associates Scheme: Guidance and planning for the national phase, 2009–2011* (HEFCE, 2009a), rings with praise for the scheme. Written by David Lammy, the then Minister of State for Higher Education and Intellectual Property, the foreword claims that the idea was a success:

> Every so often, an idea is launched and it succeeds beyond the wildest dreams. This is what has happened with Aimhigher Associates.
>
> HEFCE, 2009a: 1

According to HEFCE (2011: 2), during 2009/10, 'roughly one in every one hundred and fifty 13–18 year olds in English schools' was involved in working with university students. This scheme underwent national evaluation, which reveals a mixed picture of the scheme's successes, though the evaluation concludes that there was 'overwhelming evidence' that the programme had 'increased awareness and understanding of HE opportunities for learners' (HEFCE, 2010a: 8).

Despite this rolling out of national programmes involving student ambassadors, there has been very little research into their effects. In his review of the existing evidence of widening participation in HE, Gorard *et al.* (2007: 75) complains specifically about the lack of research into schemes involving university students, concluding that 'relatively little is known about the impact of undergraduate ambassadors, role models, mentors and tutors'. While there is mention of ambassador and similar schemes in the WP literature exploring the effectiveness of different strategies, few studies have ambassadors as their central focus. In the conclusion to their survey of the literature exploring the work of student ambassadors, Sanders and Higham (2012: 24–5) note that there is a need for research to address a range of questions about the deployment of student employees, including the importance of matching backgrounds, the relative contribution of student employees in informal and formal roles and the efficacy of different delivery models.

Wider research and evaluation of ambassador schemes in the UK has been very limited. Various studies suggest there are a number of benefits for student ambassadors, including increased confidence and capacity to learn, as well as improved positioning within the labour market (Austin and Hatt, 2005; Chilosi, 2008; Taylor, 2008). These findings reflect those of the HEFCE (2010a: 8.45) evaluation of the Student Associates scheme which identifies 'some of the most significant benefits of the programme' as being for 'the associates themselves', and again outlines benefits being to associates' 'confidence, experience and employment potential'.

Murphy (2006) provides a more critical analysis of the impact on ambassadors in her MA dissertation. She draws on Colley's (2003) analysis of engagement mentoring; Colley (2003: 530) suggests that mentoring 'entails a feminine stereotype of self-sacrifice and nurture', and that mentoring is a form of social control that modifies the behaviour of mentors and mentees. She argues that mentors are not agents of empowerment as they are themselves disempowered. Murphy warns of the 'disciplinary nature' of the ambassador scheme and suggests that student ambassadors are disempowered by it, as they are 'guided towards adopting strategies that play a role in reproducing their own subordination and social inequality' (Murphy, 2006: 69). Chilosi (2008: 7) draws on Murphy's work and comments that, as the student ambassadors in his study were themselves from deprived backgrounds, the fact that they draw from their own experience may entail their acting as 'reproducers, rather than distributors of social and cultural capital'.

Another suggestion made by Chilosi (2008: 7) is that student ambassadors' work with younger students may foster an 'authoritarian personality'. This quite interestingly relates to the findings of the HEFCE (2010a) evaluation. Ambassadors' focus on their need for authority with school pupils is evident from their explanation of their need for more help with 'classroom and behaviour management' as well as more information 'about the school curriculum'. This is a theme again picked up in a small study of Aimhigher ambassadors conducted by Ylonen (2010). It is interesting, however, to note that a high proportion of student ambassadors in Chilosi's study were studying, working or planning to work in education. This career orientation among ambassadors is reflected in the accounts of some of the participants in Taylor's (2008) study which identified being involved with tutoring programmes as facilitating access to the Postgraduate Certificate in Education (PGCE).

In her study of student ambassadors from a traditional university working with pupils in 'local' schools on tutoring programmes, Taylor (2008: 155) refutes claims that 'sameness' between ambassadors and pupils is 'encouraged and endorsed'. She identifies a 'sharpening of notions of "us" and "them"' amongst many student participants' with students contrasting their success stories against the 'educational "failures"' they find in some schools. Taylor suggests that 'social class is mobilized' in these 'constructions of the 'good student' as against the 'bad pupil'.

These studies, however, with the exception of the HEFCE report, focused on ambassador schemes from the point of view of the ambassadors themselves. Despite the focus on student voice in research with young people more generally (Fielding, 2004), there is minimal research exploring the impact of the scheme from the perspective of participating school pupils, though benefits are often claimed. The literature does generally suggest that 'HE students can provide learners with a role model from which to develop more accurate perceptions of students and challenge negative stereotypes' (Sanders and Higham, 2012: 19). Research into and evaluation of the impact of the Aimhigher Associates scheme in different regions claims that ambassadors improve understanding about accessing HE and the HE experience, and that growth in confidence among pupils is a significant outcome of their work with ambassadors (Church and Kerrigan, 2010; Thompson, 2010). Church and Kerrigan also identify improved motivation.

There is some suggestion in the literature that ambassadors can provide young people with careers guidance (HEFCE, 2011) and that they are seen as 'credible information-givers' (Hatt *et al.*, 2009: 341). Gartland *et al.* (2010) argued that student ambassadors can support the promotion of engineering messages in schools. However, the quality of this information has been questioned (Ylolen, 2010; Gartland *et al.*, 2010; Gartland, 2012/13). Slack *et al.* (2012) suggest that brief encounters with university students during visits and one-off events positions them as 'warm' sources of information about university, who can influence the decision-making particularly of disadvantaged young people, who distrust 'cold' official sources. They not only question the quality of this information, but also the focus of government on the provision of more official information as a strategy to facilitate young people's choices (Slack *et al.*, 2012).

Gartland and Paczuska (2007) suggested that student ambassadors can become trusted sources of 'hot' knowledge that is accessed and believed by younger students and that they may be a source of 'cultural capital'. While the activities carried out in the context of ambassador relationships may be relevant to the curricula of schools, colleges, or universities, they are not directed by any of them; rather, the ambassador relationship is located somewhere 'between institutions'. This social space 'between institutions' appeared to be carved out or defined independently by ambassador activity itself, which in turn frames the ambassador relationships (Gartland and Paczuska, 2007: 129). Social exchanges between student ambassadors and school and college students were considered to be important in providing younger students with valuable information about higher education, and this social 'space between institutions' appeared important, as it does not carry institutional authority. The study concluded that such relationships developed in a similar way to friendships and 'like any friendships they must be allowed to develop on the students' own terms if "trust" is to develop' (Gartland and Paczuska, 2007: 131).

There is a focus in the literature on the need for ambassador training (Ylolen, 2010; Carpenter and Kerrigan, 2009; Porter, 2010) and on the importance of recruitment and selection in identifying the best individuals to be ambassadors (Ylolen, 2010, Thompson, 2010; Sanders and Higham, 2012), and the roles they are allocated (Sanders and Higham, 2012). However, there are conclusions about other factors influencing and shaping the work of ambassadors with young people that can be drawn from the literature, albeit tentatively, and these are presented in the next two sections.

The physical location, learning contexts and focus of ambassador work

Little has been made of the physical location of ambassador work in these studies, though the physical space that ambassadors function within is significant in terms of how it facilitates or detracts from the social connection between ambassadors and school pupils. It is possible to make some inferences from the other studies discussed here about the impact of the physical location of ambassador work on the pupil–ambassador relationship.

The ambassadors in Taylor's (2008) study had worked with pupils as part of a tutoring programme called Students into Schools. Taylor describes how the university students often described these schools as 'local' and that this term had pejorative connotations. She suggests that this term was used to identify the 'specialized segregation of the "local" against the achieving, well placed and centred beacon of enlightened education that were "good" schools' (157). Some students more explicitly described their placements as being 'located within 'sink "estates" with "rough" pupils' (ibid.). This geographic, institutional and physical positioning of university students appears to have at least contributed to the sense of difference Taylor identifies as existing between pupils and ambassadors: the 'distinctions and constructions of "good" students versus "bad" pupils' (Taylor, 2008: 155). The importance of the physical location of activities is also illustrated by an evaluation of Aimhigher activity at local amateur football clubs, undertaken as part of the Higher Education through Football project. It is likely that the location of the activity contributed to the lack of conversations between students and young people about HE. In this learning context, HE is unlikely to have been seen as relevant (Carpenter and Kerrigan, 2009).

Conversely, both Taylor and the HEFCE (2010a) report make brief reference to how valuable some of their contributors viewed campus visits to be. In the HEFCE evaluation, a college tutor commented that 'the highlight of the year was a trip to the university' (7) and adds that 'this had more impact than all the rest of the activity' (44). As the study of Slack *et al.* (2012) reveals, university students in this context are often seen by pupils as reliable sources of information. Hatt *et al.* (2009) identify that there is a widely held perception among policy makers, stakeholders and participants, of the importance of summer schools. These accounts support the view that pupils attending activity physically located in universities is particularly beneficial in supporting widening participation.

The focus of ambassador work also appears to be important. The Student Associates scheme had specific foci for work with pupils. This included life at university, aspirations and progression, applying to HE, subject-specific support, and revision practice and study skills. The associates worked with groups of pupils over an extended period of time, focusing on these areas – the focus of sessions varied to some extent depending on the approach taken within the region. Both the evaluation of Aimhigher Associates (HEFCE, 2010a) and Taylor's study reveal how focusing work with pupils on access to university is potentially inappropriate with school pupils who have no existing interest in going to university, or for whom university has no immediate resonance or relevance. The HEFCE report identifies that associates found this particularly challenging. Taylor (2008: 161) even suggests that instilling the message in some school pupils that success is only achievable via going to university 'may at times serve to inscribe impossibility and failure'. Taylor outlines the danger that Aimhigher initiatives could sharpen 'the dichotomy between "achievement" and "success" and non-participation and "failure"' (Taylor, 2008: 162). The desire expressed by associates in the HEFCE report for information about qualifications and progression routes implies that they have been providing careers information to all the pupils they worked with, but that they may not have been sufficiently equipped to do so effectively, a finding supported by Ylolen (2010).

Organizational structures

I would also argue that the organizational structure within which these schemes operate and the need for WP practitioners to balance stakeholder interests is vital to any analysis. Since 2000, there have been clear and dominant discourses circulating within education that have surrounded and defined the work of WP units. Neo-liberal discourses of the marketplace and of individualism have been central and continue to dominate.

Defining the work of WP units are the layers of administrators established within universities to facilitate WP work, the employees themselves, their positioning within the HEI and their backgrounds. These WP practitioners largely operate outside of the academic faculties of universities (Burke, 2012) and are often located in marketing and recruitment departments. Their own backgrounds and beliefs are significant to the ways they work with ambassadors and school pupils; Wilkins and Burke (2013) point to how some WP professionals negotiate neo-liberal discourses dominant in policy and actively shape activities according to their own 'ethical, moral and social class commitments'. However, lines of management within universities impose constraints on WP units through the positioning of their employees and demands made on them, as well as through allocation of funding particularly in relation to recruitment targets. This inevitably influences WP practices and studies point to student ambassadors' positioning as 'marketers' of university (Ylolen, 2010; Taylor, 2008); in Ylolen's study (2010: 102) students describe a 'marketing mentality' among some ambassadors.

Aimhigher, New Labour's most extensive WP initiative in England, provided funding for many WP initiatives in universities at the time of the study that is the focus of this book (see Appendix). The dominance of aspiration-raising discourses in the work of Aimhigher and the corresponding positioning of young people as in 'deficit' has been discussed (Thomas, 2001; Burke, 2002; 2006; 2012; Archer and Yamashita, 2003; Yorke and Thomas, 2003; Watts and Bridges, 2004; Bridges, 2005). Under New Labour, Aimhigher developed increasingly prescriptive targeting criteria relating to the cohorts of young people that universities were funded to work with.

Other important stakeholders in WP work, written into the partnership working of Aimhigher, were schools. Teachers in schools have large workloads and busy timetables, so it is often difficult to engage schools with outreach projects run by universities (Gartland, 2009; HEFCE 2010a). This means that universities have to ensure that their projects appeal to teachers. Inevitably, in the league-table culture that exists, teachers' priorities are the exam results of their pupils (Williams *et al.*, 2010). This focus is different to that of the WP units with their general aspiration-raising remit.

The lack of coherence and the tensions caused by different stakeholder foci in outreach work are evident in the evaluation of the Aimghigher Associates (HEFCE, 2010a). Some schools 'specifically selected lower ability groups with a view to improving attainment levels' (HEFCE, 2010a: 64). The different agendas of schools and organizers appear to have led to a lack of consensus both about learner selection for the Student Associates scheme and with regard to the purpose of the programme as a whole (HEFCE, 2010a; Sanders and Higham, 2012). Indeed, a problem for the Aimhigher Associates scheme as a whole appears to have been a lack of clarity over the project aims. Reflecting on these constraints, the suggestion that the relationship between ambassadors and pupils can exist in a social space that is separate from these institutions understates the impact of institution and stakeholder agendas on shaping these relationships (Gartland and Paczuska, 2007).

Summarizing the research evidence

A few tentative conclusions can be drawn from the limited research into the work of ambassadors. There are clear benefits for ambassadors who undertake work with school pupils on these schemes. Benefits to school pupils are also claimed in terms of their raised aspirations, knowledge of HE and cultural capital more generally, although these are not coherently represented through research undertaken so far, very little of which draws on the views of pupils themselves.

Ambassadors often appear to feel responsible for the discipline and control of pupils. This positioning seems to be compounded by the location of work, with ambassadors undertaking some sort of tutoring work in schools, and by ambassadors' own interests in becoming teachers themselves. Visits to university were viewed by some contributors as more valuable than time spent by ambassadors in school. Given the potential significance of the physical locations and learning contexts within which they work, I suggest that it is useful to identify ways to conceptualize these differences. Theories from work-based learning, including theories of 'formal' and 'informal' learning (Eraut, 2000; Beckett and Hager, 2002; Hodkinson and Hodkinson, 2001; European Commission, 2001; Colley *et al.*, 2003), are helpful for this process, as the balance of informal and formal attributes in these learning contexts 'inevitably changes the nature of the learning' (Colley, 2005: 31).

For pupils who are hostile to the idea of going to university and for those who do not have the academic credentials to progress to university, ambassadors' focus on the potential benefits of going to university seems inappropriate and even undermining of some young people. This reflects other research that highlights the importance of students' biographies and learner identities to their progression in education. (Ball *et al.*, 2000; Brooks, 2003a; Evans, 2006; Reay *et al.*, 2009).

Ambassadors may not be informed enough to provide pupils with the careers advice that pupils request, and they may 'reproduce' rather than 'distribute' social and cultural capital as they, in some instances, come from deprived backgrounds themselves and focus on their own experiences. Alternatively, students' own different, contrasting backgrounds can serve to entrench a sense of difference between ambassadors and school pupils. It is also suggested that ambassadors' familiarity with the school they visit and shared interests held with pupils may be as significant as matching backgrounds of pupils and learners. Whether it is important to match backgrounds, which aspects of students' and pupils' identities to match and what the benefits are remains unclear.

The foci of policy and the interests of the different stakeholders involved in ambassador work are significant in shaping the work that takes place between school pupils and ambassadors, and this in turn appears to shape the relationships that are facilitated.

Mentoring and role models

Literature on youth mentoring focuses almost exclusively on the impact and nature of one-to-one relationships between adults, adults and youths, or adolescents and children. While these relationships are inevitably different to those between student ambassadors and school pupils, as they are generally more intense and longer in duration, there are many similarities and parallels. A brief look at the research exploring these mentoring relationships is revealing.

Formal youth-mentoring schemes often have specific outcomes in mind. Much of the literature on youth mentoring comes from the United States, where mentoring first developed as 'a formal response to social exclusion and social welfare problems' (Newburn and Shiner, 2006: 23). Keller (2007) charts the development of 'prevention-orientated approaches' that became prominent as state involvement in social issues increased during the twentieth century; these tended to be 'categorical in nature, attempting to identify and reduce risk factors associated with specific problem behaviours' (Catalano *et al.*, 2002: 26). A different approach to mentoring has been through positive youth development programmes (Roth and Brooks-Gunn, 2003, in Keller, 2007: 26) aiming to 'promote the development of life skills and foster the talents and abilities of youth'. There are also programmes that focus on issues within communities and social trends affecting young people, such as single parents and labour patterns that have resulted in limited contact between children and adults, with many young people 'spending their time in age-graded settings or home alone' (ibid.).

Government funding for student ambassador schemes and their development in UK universities fit into this wider context and history. Colley (2003) outlines how, by the mid-1990s, youth mentoring had taken root in the UK; at this time, the Mentoring Action Project (MAP) was undertaken by the Institute of Careers Guidance (ICG), and the National Mentoring Network (NMN) was established. When New Labour came to power in 1997, mentoring schemes expanded further, with the NMN receiving increasingly large bursaries. In 1998, the House of Commons Select Committee on Disaffected Children outlined that all programmes focusing on disaffection should include mentoring. Colley (2003: 522) explains how mentoring became 'a standard ingredient in the recipe of almost every major new policy initiative, including prevention of school truancy and drop out from post compulsory education and training'.

A number of theoretical frameworks have been devised to explain the influence of mentoring relationships; Rhodes (2002) has been influential. Rhodes suggests that mentors promote youth development in three ways: 'a) enhancing skills and emotional well being, b) improving cognitive skills through instruction and dialogue and c) fostering identity development by serving as a role model' (cited in Keller, 2007: 37). However, the literature exploring formal youth mentoring programmes does not paint an overwhelmingly positive picture; benefits are not categorically clear (McPartland and Nettles, 1991; Abbot et al., 1997; Aseltine et al., 2000; Tierney and Grossman, 2000; DuBois et al., 2002; Sipe, 2002; Colley, 2003). As Blinn-Pike (2007: 183) suggests, mentoring has 'come to be regarded as more complex than first recognised' and it is becoming clear that 'mentoring alone will not help high-risk youth solve all their problems'.

Colley's (2003) study of 'engagement mentoring' explores a UK programme using university-student mentors to 'improve' the attitudes to work of socially excluded young people. She expresses concern about such programmes because their narrow emphasis reflects the needs of the economy rather than the needs and development of individuals. In a way that resonates with the structural issues raised in the HEFCE (2010a) evaluation of the Student Associates scheme, Colley points out the difficulties encountered by mentors attempting to promote routes into employment supported by organizations but rejected by individual young people. It seems that placing people in a relationship where one is prescribed the task of encouraging others to embark on work or study that they are not interested in and even opposed to is beset by difficulties.

Specifically educationally oriented mentoring programmes have had various ambitions: Blinn-Pike (2007: 170) identifies common targets as being 'improvements in academic grades, attitudes toward school, school behaviour, absenteeism and preparation for college'. There are various studies that reveal positive outcomes for pupils (Cave and Quint, 1990; Johnson, 1997; 1999; Starks, 2002). Cave and Quint (1990: 170) suggest that mentoring can support and contribute to the benefits pupils receive from other career services. However, positive outcomes are not guaranteed; one study conducted by Abbot *et al.* (1997) actually appears to suggest that a mentoring programme targeting 22 boys from mother-headed households impacted negatively on their grade point average.

There has been little focus in research on naturally occurring mentoring relationships (Spencer, 2007). Concerns have been expressed about there being fewer opportunities in recent years for naturally occurring mentoring relationships to develop, especially in urban centres (Spencer, 2007). However, Spencer outlines the findings of research studies such as those of Zimmerman *et al.* (2002) and Chen *et al.* (2003) which suggest that, despite constraints on opportunities to meet adults, naturally occurring mentoring relationships continue to be a part of many young people's lives. Naturally occurring mentors include extended family members and non-familial adults in professional roles and people connected to young people informally such as boyfriends and girlfriends of family members or friends' parents or siblings (Spencer, 2007: 101). A range of benefits for young people is attributed to naturally occurring mentoring relationships with adults. The social support provided is thought to be valuable for young people. This social support has been identified to include instrumental support, emotional support and companionship support (Cohen and Willis, 1985). However, there is also research to suggest that mentors drawn from families can actually serve as negative role models that can act to compound difficulties faced by young people (Sanchez *et al.*, 2006).

While programme mentoring rarely allows for matching the backgrounds of mentees and mentors because of the shortage of available volunteers, research into naturally occurring mentoring relationships suggests that the mentors young people choose are like themselves in terms of gender, racial, ethnic and class backgrounds (Spencer, 2007; Klaw and Rhodes, 1995; Sanchez and Reyes, 1999; Chen *et al.*, 2003). There is a body of research suggesting that matching backgrounds of mentors and mentees is beneficial to the relationship (Klaw and Rhodes, 1995; Sanchez and Colon,

2005; Cavell *et al.*, 2002; Jackson *et al.*, 1996; Kalbfleish and Davies, 1991; Sanchez and Reyes, 1999; Chen *et al.*, 2003; Ensher and Murphy, 1997). However, these studies do not conclude that matching backgrounds in mentoring relationships is necessarily better in all respects. For instance, Ensher and Murphy (1997) found that mentees in cross-race matches are as satisfied as those in same-race matches and saw themselves as like their mentors in other ways (Liang and Grossman, 2007). There have been various studies to suggest that similarity of interests and attitudes are more important than demographic similarity (Ensher *et al.*, 2002; Ensher and Murphy, 1997; Grossman and Rhodes, 2002). Perceived similarity between themselves and mentors appears to be important to young people, but similarity 'may be indicated by qualities such as shared interests and geographic proximity' (Liang and Grossman, 2007: 251). There appear to be 'complex interactions between demographic characteristics and multiple aspects of a youth's identity' (Liang and Grossman, 2007: 251).

The findings of research related specifically to STEM suggest that mentors can support women progressing in STEM careers and subjects (Ragins and Cotton, 1999; Ambrose, 1997; Etkowit *et al.*, 2000). Chesler and Chesler (2002: 50) claim that STEM mentoring can 'have a significant impact on careers and lives'. Packard (1999) suggests that young people prefer to work with mentors who they perceive to be similar to themselves. Packard suggests that, for women studying science subjects, a 'composite mentor' – where students have a range of diverse mentors – is beneficial. Her reasoning is that, as there are so few female mentor images in the field, having a diverse set of mentors in this way would 'help women make use of the available images in their environment' (Packard, 1999: 52). This lack of 'role models' in STEM is also evident in the school environment, as there are few female teachers of these subjects (Cano *et al.*, 2001).

To combat this lack of 'role models' in schools, STEMNET's STEM ambassadors scheme has massively extended over the last decade in the UK, and now has approximately 28,000 volunteers. The STEMNET website highlights these ambassadors' work as role models: 'STEM Ambassadors are people from STEM backgrounds who volunteer as inspiring role models for young people'. The organization currently receives funding from the BIS, the Department for Education (DfE) and The Gatsby Charitable Foundation. An evaluation undertaken at STEMNET (Straw *et al.*, 2011) explored the impact of the work of STEM ambassadors with school pupils. Survey data from this evaluation suggest that STEM ambassadors do promote 'increased engagement and interest in STEM subjects', 'increased knowledge and understanding of STEM subjects' as well as 'increased awareness of the

STEM employment and careers options' and 'study options available' (Straw *et al.*, 2011: 6). The report also argues that, given that almost half of pupils surveyed had only seen an ambassador once, 'much more could be achieved with ongoing and sustained contact' (Straw *et al.*, 2011: 6).

While theory and research relating to role models is no longer widely drawn on academically, the assumption that adults working with young people are automatically 'role models' is ubiquitous. There are, however, critiques of assumptions that adults working with pupils are seen as 'role models', and in some instances these question whether presenting pupils with successful role models is always beneficial. Individuals can only become role models for young people if young people choose to emulate them (Marx and Roman, 2002). The evaluation of STEM ambassadors does not take account of the learning that is taking place within these relationships or of the interplay of learning contexts and identities and how this affects processes of identification between school pupils and ambassadors. That STEM ambassadors are seen as 'role models' by pupils is, then, largely assumed. Delgado (1991) provides a powerful and scathing criticism of the way in which successful black professionals like himself are presented as 'role models' in schools in order to encourage young people from deprived areas to aspire to professions that they are structurally excluded from. However, while the critique is important, the literature does suggest that there can be real benefits to providing young people with role models. Marx and Roman (2002: 1184) go on to suggest that there are 'some important benefits to comparing oneself to a similar and outstanding other': role models can be seen as aspirational (Lockwood and Kunda, 1997; Tesser, 1986): to enhance self-evaluations and motivation (Blanton *et al.*, 2000; Collins, 1996; Major *et al.*, 1993; Taylor and Lobel, 1989) and to guide students' academic aspirations (Hackett, 1985; Lockwood and Kunda, 1997).

Chapter 4
Analysing policy and practice: a multi-stranded approach

Ball (1994: 14) stresses the need for a range of approaches when analysing policy: 'a toolbox of diverse concepts and theories – an applied sociology rather than a pure one'. He suggests that a challenge for research attempting to consider the 'macro' level of policy with the 'micro' level of practice is to be able to relate them; 'to look for the iterations embedded within chaos'. The approach in this study attempts to relate the micro-level interchanges between ambassadors and pupils to the macro level of HE and the wider education policy context. Diverse concepts and theories, including Foucauldian discourse analysis and the theories of post-structuralists, especially Judith Butler, as well as a 'toolbox' that draws from practices in social psychology, ethnography and grounded theory, have facilitated this approach.

A discursive strategy

In Foucault's view, knowledge operates in institutional settings through institutional 'apparatus' and 'technologies' (Hall, 1997: 27). In *Discipline and Punish*, Foucault (1977) explores how what is viewed as being 'known' in a particular time period impacts on how we treat criminals: how they are regulated, controlled and punished. He considers how institutions are powerful in developing salient knowledge and implementing control. Indeed, Foucault sees the whole of society exerting this control 'because institutions improvise, cite and circulate discursive frames and coterminous technologies that render subjects in relations of power' (Youdell, 2006: 571). This occurs through a regime of truth:

> Each society has its regime of truth, its 'general politics' of truth; that is, the types of discourse which it accepts and makes function as true, the mechanisms and instances which enable one to distinguish true and false statements, the means by which each is sanctioned ... the status of those who are charged with saying what counts as true.
>
> Foucault, 1980, in Hall, 1997: 131

Analysing policy and practice: a multi-stranded approach

Ball (1994: 22) explains that the state is 'decentred' in this; 'the state is ... the product of discourse, a point in the diagram of power'. However, Ball also identifies the existence of 'dominant' discourses that currently circulate within social policy and that increasingly develop in hegemony through policy; he identifies neo-liberalism and 'management theory' as two such dominant regimes of truth (Ball, 1994: 24).

In recent years, social psychology has embraced Foucauldian discourse analysis (Hollway, 1984; Parker, 1992; Willig, 2001; Wetherell and Potter, 1992; Wetherell, 1998; Wetherell *et al.*, 2001a and 2001b). The emphasis in social psychology is on how people's positions, thoughts and actions are presented as being delineated (the extent of this delineation is much debated) by available discursive resources. Parker (1992: 245) defines discourses in Foucauldian terms as 'sets of statements that construct objects and an array of subject positions'. The subject of these Foucauldian studies is decentred 'into numerous subject positions' (Anderson, 2003: xvi). Edley (2001: 210) describes the 'concept of subject positions' as 'central' to discursive psychology, suggesting that this 'connects the wider discourses and interpretive repertoires to the social construction of particular selves'.

Positioning theory provides a useful analytical tool to examine how discourses circulating in HEIs develop regimes of truth, how knowledge operates through the apparatus and technologies of these institutions and regimes of truth and how these regimes of truth define and constrain the subject positions available to employees, including student ambassadors and the young people that they work with. In my exploration of the relationships between ambassadors and school pupils, the positioning of student ambassadors and how this affects the relationship they develop with pupils has been of central importance. I have also considered ways in which ambassadors are positioned by the HEIs and other institutions in which they work, exploring ambassadors' understanding of their positioning and their perspectives about how this affects their relationships with pupils, as well as the perspectives of the pupils themselves.

This entailed examining the discourses associated with the positioning of ambassadors and of members of WP staff. Ball (1994: 19) suggests that staff responses to policy texts have to be 'creative', as policies 'do not tell you what to do' but may set 'particular goals' or 'narrow the range of options available'. In this interpretative process of how policy is implemented 'on the ground', the individual outlook and histories of employees within institutions and their positioning within the institutions themselves are important.

This approach has been useful in exploring the way that ambassadors were positioned, but further analytic tools were required to explore the relationships that developed and how positioning affected these relationships. The approach used in discursive psychology, drawing on conversation analysis, with its focus on 'how people use discursive resources in order to achieve interpersonal objectives in social interaction' (Willig, 2001: 91), has also been relevant, both in terms of how ambassadors and pupils talk to each other and with regard to how ambassadors and younger students talk about themselves and each other. Also relevant has been the 'inclusive' understanding of discourse proposed by social postmodernists, particularly Laclau and Mouffe, where 'discourse is equated with the social or with human meaning making processes in general' and 'discourse includes both linguistic and non-linguistic elements' (Wetherell 1998: 392).

I have drawn on a wide range of data sources, including interview data from meetings with pupils and ambassadors and observation data from different interventions. I have also examined discourses about ambassadors circulating within HEIs. To do this, I have listened to talk within HEI contexts and explored the history or, in Foucauldian terms, the 'genealogy' of these discourses.

Much WP intervention is based on the premise that many students will self-exclude from HE based on the cultural norms of students from lower socio-economic groups and groups without a family background in HE . The aim of many WP projects is to 'raise aspirations' (Burke, 2012) by providing students with information and insight into HE, so they can make an informed rational decision. This vision of how students behave is rooted in a liberal-humanist understanding of individuals who are at liberty to 'choose' what they want to do and what sort of person they want to be. Clearly, Foucault's conception of a subject as constituted through discourse challenges this.

However, if discourses are viewed as totally controlling how people act, that presents individual subjects as being without agency. If this is the case, how does change happen? Wetherell and Edley (1999) and Billig (1991) all identify the subject as being both produced by and producers of language. This paradoxical relationship carries with it the suggestion that speakers, though constrained by language and regimes of truth imposed through available discourses, also have agency as the producers of language in any given context. Positioning theorists also argue against the conception of subjects as being without agency; they move beyond the contexts of particular conversations to talk more generally about individual agency in identity construction:

> Positioning theorists argue against a wholly agentless sense of master discourses in which identity construction is constrained by

> a restrictive set of subject positions available. Instead they claim that people may resist, negotiate, modify or refuse positions, thus preserving individual agency in identity construction.
>
> Benwell and Stokoe, 2006: 43

Judith Butler has perhaps most effectively theorized a subject that is both constrained and agentic. She suggests that our gendered identity is performed and that the 'performative character' of gender simultaneously provides the possibility of challenging hegemonic meanings:

> If gender attributes ... are not expressive but performative, then these attributes effectively constitute the identity they are said to express or reveal ... That gender reality is created through sustained social performances means that the very notions of an essential sex and a true or abiding masculinity or femininity are also constituted as part of the strategy that conceals gender's performative character and the performative possibilities for proliferating gender configurations outside the restricting frames of masculinist domination.
>
> Butler, 1990: 140–1

Butler's ideas of the performative nature of social interactions and the potential for these to be reproductive but also transformative has been of interest in my analysis of learning contexts and ambassador–pupil interactions.

The researcher

I am aware that my own positioning will inevitably influence the picture that has emerged from my study. I cannot be outside the confines of available discourses; I am constructed and positioned by them and this inevitably impacts on the research I undertake. I am aware that my own biography has contributed to my understandings. My extensive work as a teacher and researcher in the geographical and conceptual area within which the study is based has inevitably honed my focus on, and practical understanding of, learning and teaching issues in these contexts. This background underpins the theoretical work in my research and positions me, providing the particular critical lens that informs my approach. However, in my analysis I do not intend to 'reauthorize' myself through 'telling and confession' (Skeggs, 2003: 363). As Skeggs suggests, it is in the 'practice' of research that reflexivity is important.

As a white middle-class professional woman, my background distances me from the participants of this study, most of whom are young, many are male and a large number are black. Participants are likely to respond to my presence in particular ways. There has been much discussion of the importance of matching the background of interviewers and interviewees (Skeggs, 1997; Maynard, 1998). It may be the case that participants of a study are more likely to discuss details of their personal lives with researchers who are similar to them and could better understand their perspectives. However, this proximity may be conceived as equally problematic as such proximity may make participants feel positioned in particular ways and so lead to a reluctance to articulate views that are contrary to expectations. Walkerdine *et al.* (2003: 191) critique any claims that it is 'possible for researchers and subjects to be equal', suggesting that this is a 'fantasy'. As they point out, 'looking at who reveals what to whom involves complicated plays of poser'. Gerson and Horowitz (2003: 212) identify certain advantages associated with having a social identity 'outside the usual social order' and note that this provides the opportunity to 'gain a unique perspective'. What is perhaps important is that different researchers are likely to discover different insights from the same participants, but that these insights are no less useful for being different.

The institutions

The two institutions[1] where the study is based illustrate clearly the stratification within the HE sector. I was keen to explore the impact that this stratification had on ambassador work and on the relationship between ambassadors and school pupils. Both universities were set up in the nineteenth century and are geographically very close together, but they are very different in terms of their student body and subject offer. Royal is part of the 'golden triangle' of elite British institutions and has always provided an elite education to a select group of initially male students from affluent backgrounds. In contrast, Bankside was originally set up as a polytechnic institute for the poorer classes with a focus on education to equip participants for 'the workplace'. The institute was for local people and admitted both men and women from the outset.

History is still very much embedded in the practices of both institutions. Bankside still recruits from the local area, whereas Royal has a wider national recruitment strategy, selecting only the most well-qualified applicants. Bankside, both in the past and in the present, has vocational education as its focus and continues to cater for local people, though the university also targets the lucrative 'market' of international students. This vocational focus serves to distance Bankside from more elite universities, with their more singular emphasis on academic excellence. Royal targets only the 'academic elite', both in the UK and internationally, as the institution works to develop its identity within the globalized knowledge economy. A comparison of the student bodies at the two institutions also illustrates the differences in the types of students the universities recruit. There are differences in the ethnicity of students at the two institutions, with a high proportion of Asian students at Royal and more Black African students at Bankside; there is also a slightly higher proportion of white students at Royal, but what is most striking is the difference between the student bodies in terms of class. Under 5 per cent of the full-time undergraduate population at Bankside is drawn from private schools and 40 per cent is from socio-economic groups 4–7,[2] while at Royal well over a quarter of the full-time undergraduate population is drawn from private schools with less than a quarter coming from socio-economic groups 4–7 (HESA, 2009/10).

Royal offers engineering subjects, but the offer only extends to postgraduate students; undergraduate courses have been withdrawn. This closure of undergraduate engineering courses reflects national problems that have faced engineering subjects, since it has sometimes been seen as more cost-effective for departments at elite institutions to focus on research, as the cost of teaching engineering students is high and the demand for courses relatively low. However, medicine at Royal recruits high numbers of undergraduate students. Royal is one of the largest centres for healthcare education in Europe, and the School of Medicine has over 2,000 undergraduate students. This focus on medicine is linked to Royal's status as an elite institution. The university was founded by professional middle-class men for others who would also enter into 'the professions'; medicine ranks among the oldest and most established of these. Medicine at Royal draws predominantly on students from middle-class backgrounds; just over a quarter of students were drawn from socio-economic groups 4–8 (HESA, 2009/10). Women outnumber men on undergraduate courses in medicine.

Bankside was opened in Victorian times to fill a skills gap identified by the political elite. At the time, both Liberals and Tories were 'convinced of the urgent need in this country for technical and commercial education' (university website) – a familiar preoccupation which resonates with contemporary political discourses. Bankside was to provide such 'industrial skills' to local people. Engineering at Bankside is now strongly represented: students studying engineering and related subjects make up almost a quarter of the student population at the university. The focus on links with industry is still evident at Bankside, and publicity about courses stresses this and the employability of students who attain degrees there. As with the university as a whole, ethnic minority groups and lower socio-economic groups were well represented in 2009/10: 76.4 per cent of engineering and technology students from Bankside were from socio-economic groups 4–8. However, reflecting national and international patterns, women are still strongly under-represented in these subject areas (HESA, 2009/10).

STEM widening participation activities

Two particular schemes responsible for a range of outreach activities were running at the universities at the time of the study. At Bankside, there was a HEFCE-funded project, the Accessing Engineering Project (AEP). The aim of this was to engage more diverse groups in engineering (Appendix), including ethnically diverse pupils, pupils with no family history of HE and girls. STEM student ambassadors were used extensively during AEP outreach activities. These ambassadors were trained and recruited by both the central WP unit at Bankside and by AEP project organizers, some of whom were subject specialists. At Royal, the Medical Access Scheme (MAS), an extended six-year degree programme, has been designed to enable students with lower A-level grades than their peers on the standard five-year programme to progress to careers in medicine. Student ambassadors, themselves on the degree programme, were employed to raise awareness about the MAS and to encourage young people from local schools to aspire to study medicine. Unlike the WP coordinators in the central WP unit at the university, the outreach manager, responsible for these ambassadors' training, was a subject specialist and a qualified doctor. Medical students on the programme were recruited from London boroughs from schools with 'poor academic records and poor progression rates into HE' (outreach manager, Royal).

The central WP units at both universities also organized outreach activity. There was a large centrally based programme of WP activities at Bankside that utilized student ambassadors and targeted a range of inner-city London schools. At Royal, the central WP unit organized a range of activities. WP activities at both universities were funded by the institutions themselves and by Aimhigher. At the time of the study, Royal funded some WP activity itself, but only with gifted and talented (G&T) pupils.

I have observed and held informal group conversations/focus groups at various activities. As well as observing the activities themselves, I observed the ambassador recruitment processes at both universities, attended a meeting with Bankside staff and other stakeholders, discussing how to improve the ambassador recruitment process, and held interviews/conversations with key members of staff at both universities and Aimhigher about their work. I also talked to organizers and teachers at events and activities. Over the course of the study, informal conversations/focus groups were conducted with 41 pupils and 16 student ambassadors at Royal and 71 pupils and 16 ambassadors at Bankside.

Bankside

Several one-day events involving Bankside students, all of them part of the AEP, were observed. Events observed included a STEM day, a careers afternoon and an event (Train Tracks) at an engineering-related workplace. At the STEM day, pupils were at Bankside University and were involved in designing cars that they had to present to a group of 'Dragons' – professional engineers. The careers afternoon was organized by a London borough to promote careers in science and engineering, and was again held at a university site (the AEP was one of a number of STEM organizations contributing to this event). Train Tracks was held at offices in central London, and was a collaboration between the AEP and Rail Links; pupils were involved in practical engineering activities working alongside ambassadors and Rail Links employees (STEM ambassadors).

Another AEP activity observed was an engineering camp run by the Engineers Trust for the AEP, based at a rural university campus. I stayed two days on the camp, observing and talking to participants and organizers. During the engineering camp, pupils were involved in different projects: groups of pupils designed, made and presented various products, including a model eco house and a model aeroplane. Student ambassadors, called 'supervisors', were closely involved in managing pupils as well as working with them.

I attended three days of the five-day Bankside Aimhigher summer school, observing pupils on an engineering course which ran in the morning (and was supported by the AEP) and during extracurricular sessions in the afternoon. The focus during morning sessions was robotics. Pupils discussed the ethics of using robots, built and programmed robots, attended a lecture about robotics and worked on a presentation for the rest of the summer school about robotics. In the afternoons, extracurricular courses ran that pupils could select to attend; these were again supported by ambassadors.

Another activity observed was a maths workshop running after school in a classroom at a girls' school in south east London. The workshops were organized by the school and supported by Aimhigher. The workshops included both Royal and Bankside students, but the Royal student ambassadors did not attend regularly and were not present during the sessions I observed. The Year 11 pupils attending these workshops were all borderline C/D grade candidates; it was hoped that the individual support offered by ambassadors would help them achieve a C grade. I attended four weekly sessions.

Royal
I attended two one-day events organized as part of outreach for the MAS. Both were held at the university. One medical day was practical and introduced pupils to some medical science: in the morning, pupils collected their own DNA and discussed the benefits and disadvantages of stem cell research, and in the afternoon pupils were given a tour of the campus and analysed a collection of diseased organs. The other event attended was an afternoon about 'key skills for medical professionals'. Ambassadors worked with pupils to brainstorm skills and to facilitate practical exercises.

I spent one day attending the four-day G&T summer school based at Royal. Activities for the first two days were based around medicine, health sciences and arts and humanities. Pupils interested in science attended sessions focused on medicine: a heart dissection, a lecture on 'The Heart' and a clinical skills workshop. There were also general introductions to courses and to the university that were attended by all pupils. On the day attended, sessions were of more general interest. Pupils worked in non-subject-specific groups to prepare a presentation for the whole group, to be given the following day. I followed a group of pupils who had attended sessions focused on medicine as they worked with ambassadors preparing their presentations.

The participants: gender, class and ethnicity

Organizers from the WP units at each HEI and those specifically linked to the AEP and MAS contributed considerably to the study. My initial intention was to work collaboratively with WP professionals, to ensure that the study had practical outcomes and was of use to them in their professional practice. However, this was made impossible by the repercussions of the financial crisis from 2008 and rapidly shifting funding landscape, with several key members of staff changing jobs.

Studies that focus on pupils' experiences and learning are mostly absent from the literature about student ambassadors; an aim of this study was to address this gap. However, given the limited extent of pupils' contact with ambassadors, there was little opportunity for me to involve pupils in the research process; individual pupils could only offer fragmentary, brief insights into their experiences of particular contexts. With the fragments collected, I have been able to build a picture of ambassador–pupil interaction across contexts. It is, however, important to note the pupils' voices represented here (Fielding, 2004). This study may do little to serve the 'interests of students who are least well-served' (Silva, 2001: 98); pupils heard during this study have established learner identities, are generally successful in and engaged with education. I hope, however, to reveal 'power relations which create voices' and the 'voices created by the pedagogies' (Arnot and Reay, 2007: 312). This approach should provide insight for those routinely excluded as well as for pupils who are at the centre of the study.

The labelling of pupils participating in WP schemes is problematic (David *et al.*, 2010; Stuart, 2006; Burke, 2012). Stuart (2006: 163) identifies a range of terminology used 'often ... interchangeably', including 'non-traditional', 'under-represented', 'working class', 'widening participation students', and 'first generation entrants'. Stuart suggests that this can confuse the findings of research, as findings may not reflect the 'experience of specific groups' because these are not clearly defined. The range of schemes considered in this study contributes to this complexity, with groups under-represented in HE meaning different things for STEM subjects and in HE more generally. Women constitute the most significantly under-represented group in engineering at HE, while this is not the case in HE more generally or in medicine. The gendered identities of participants and how this intersects with other aspects of identity (Crenshaw, 1989; Mirza, 2008; Morley, 2012) have been important foci of this study. I consulted similar numbers of male and female ambassadors at events. The recorded focus groups/conversations included more girls than boys. This was in part due to larger numbers of girls than boys attending some events, with the maths workshops taking place in a girls' school, and partly because more girls volunteered to contribute. During informal conversations where I took notes I talked to similar numbers of male and female pupils.

I did not gather precise information about pupils' socio-economic or educational backgrounds, as I decided that it would be unethical to ask school pupils directly for details about their socio-economic status and family background. The opportunity and time frames for such conversations were generally limited. Focus groups/conversations were frequently held with other young people they scarcely knew, and I was aware that they could feel uncomfortable responding to direct questioning about, for example, their family's work and educational backgrounds. However, pupils often volunteered information about what members of their family did when discussing their own plans. I asked all student ambassadors if their parents had been to university in order to provide an indication of their social class, although this does not preclude them from being from relatively middle-class backgrounds (Stuart, 2006). The schools and districts from which pupils came, however, provide some insight into their socio-economic and class status. All the AEP and one of the MAS activities were targeted at south east London state schools from 'deprived' boroughs, with extremely low participation rates in HE, according to the Index of Multiple Deprivation (IMD, 2004). The maths workshops were also attended by pupils from a south east London school within one of these boroughs. The G&T summer school at Royal drew from a broader cohort of state schools across London. The 'voluntary' basis of pupils' attendance at events however, means that pupils from the lowest socio-economic groups who are most dramatically under-represented in HE (Galindo-Rueda et al., 2004; Sanders, 2006) are unlikely to be well represented in this study, as they are unlikely to have volunteered or been approached to volunteer by teachers.

While the pupils and ambassadors in this study may not represent the most socio-economically disadvantaged groups, the indicators gathered about their backgrounds suggest that, with the clear exception of those involved on the G&T summer school, they are predominantly from working-class and lower-middle-class backgrounds (Brooks, 2003a). Class is a broad term incorporating other group definitions, and I follow Skeggs (1997: 4) in believing that class is important: 'The historical generation of classed categorizations provides discursive frameworks which enable, legitimate and map onto material inequalities.' The pupils and ambassadors were ethnically diverse, the largest group being Black African. Pattillo-McCoy (1999: 1) suggests that, in the USA, the black middle classes are hardly visible because of the focus on 'poor urban ghettos'. This is also true of London, and there are middle-class black families living in areas of south east London whose children attend local schools. However, as Reay et al. (2005) argue, for some young people, while their parents may have been 'middle class in their country of origin', the process of migration detracts from the social and cultural capital usually gained from a middle-class background.

Bankside

The AEP selected schools located in the London boroughs of Southwark, Lambeth, Lewisham, Tower Hamlets and Newham for events and activities using Bankside ambassadors. These were included among the most deprived boroughs in England (IMD, 2004), where participation rates in higher education were among the lowest. The reason for this was to reach first generation HE applicants and cohorts of minority-ethnic pupils currently under-represented in engineering and STEM subjects, as well as to reach girls, as there are a number of single sex schools within these boroughs.

The AEP took an inclusive approach to pupils by presenting whole year groups with STEM-related activities. However, many of the activities discussed here involved smaller groups of pupils visiting universities or other sites. Schools selected pupils to attend. The teachers spoken to during events had asked for volunteers to attend, and pupils had been allocated places on a first-come, first-served basis. Several teachers explained that they had encouraged particular pupils who they thought would be interested to attend. The objective for all activities during the engineering project was that 50 per cent of all participants would be female. This meant that, despite the gender imbalance in engineering more generally, girls were well represented during activities considered in this study.

Royal
By contrast, the approach taken by Royal to WP work in London schools observed for this study focused exclusively on G&T cohorts. The summer school was attended by G&T pupils from a wide range of state schools and appeared to attract mostly middle-class pupils. This summer school was supported by ambassadors from various undergraduate and postgraduate courses.

Schools targeted through outreach activities for MAS, like the AEP, included those in the deprived boroughs in south east London. The medical skills day and afternoon, however, was attended by G&T pupils. Student ambassadors supporting these events were medical students themselves, studying on the MAS, and separate to ambassadors employed by the central WP unit at the university.

Collecting the data: an ethnographic approach

The individual activities did not last long enough for me to be embedded within them as in traditional ethnography (Mac an Ghaill, 1994; Skeggs, 1997; Hey, 1997), though I was, for two years, firmly embedded within the work of ambassadors across different activities in the same geographic area. However, participant observation, a strategy drawn from ethnography, provided the opportunity to both talk with ambassadors and pupils and to see them working together, as well as to capture linguistic and paralinguistic details at events. This approach enabled me to consider the learning contexts in which ambassadors and pupils were placed, the ways ambassadors were asked to work and the interaction between pupils and ambassadors. Notes made about the atmosphere during events, about the physical interaction between ambassadors and about the physical space within which activities were held, have furnished important details for the analysis. Using participant observation enabled the exploration of 'discourses and their subjectivating effects' and of 'the social process of subjective re/formation' occurring within the relationships between ambassadors and pupils (Youdell, 2006: 513).

During the one-day events, most conversations with pupils were very brief, but there were opportunities to hold longer conversations/focus groups during lunch and other breaks during the medical skills day and summer schools. Student ambassadors were usually prepared to stay behind, indeed in most instances talking to me meant that they did not have to tidy up at the end of the event, so it was viewed positively. In one instance they were even paid extra for their time by organizers.

Given the time constraints, focus groups/group conversations with pupils and paired interviews or focus groups/group conversations with ambassadors were the most propitious way to proceed with data collection. Working with others enriched the accounts of pupils and ambassadors, who were able to build on the contributions of their peers where they may have struggled to think of much to say about their relationships with each other on an individual basis. This negotiation of accounts will inevitably have influenced what was said. Gaskell (2000: 46) suggests that in group contexts there is 'the development of a shared identity' which is 'captured in the self description "we"'. I was interested in the process of this negotiation within groups of ambassadors and pupils and wanted to explore the conversations that led to this shared identity in terms of their relationships with each other.

Organizers agreed to focus groups/conversations taking place during activities, though none of these were organized in advance; organizers had little contact with schools before events and I was keen not to present an additional administrative burden. During the activities, where time frames permitted, I approached pupils directly to ask if they would be prepared to talk to me during their break/lunch break, or asked for volunteers at the start of events. The one-day events were hectic and activities filled the time (often even including lunch). The approach I developed as a result was to circulate and sit with pupils while they were working on tasks and, if they were happy to talk, to have conversations with them as they worked. I did also sometimes work with pupils while we talked, for example cutting and gluing, or discussing information for presentations. These conversations lasted for anything from five to fifteen minutes.

At the start of all conversations I briefly explained my research. I spoke generally about my focus being on the activity they had been involved in and what ambassadors contribute to it. The tone of this introduction was important, as I wanted to avoid pupils feeling that I was asking them to evaluate the ambassadors and make judgments about their contributions – I rather wanted to elicit stories about their day. When introducing the research to ambassadors I explained my focus in more detail; I felt that this awareness would lead to 'more analytical reflection' about their work (Robson, 2002:

317). Ambassadors seemed very happy to talk and to share their experiences; most were highly committed and enthusiastic about their work and were keen to share stories about their interactions with pupils.

During longer focus groups/conversations, an interview schedule was used, though not prescriptively. The approach taken was to provide a context within which pupils and ambassadors could tell their own stories about their experiences and their work together. The hope was to enable participants to construct and co-construct accounts with their own emphasis and focus rather than one set by the interviewer. This approach allows a 'fuller representation of the interviewee's perspectives' (Mason, 1998: 42). Open-ended questions were also followed up with questions relating specifically to what participants said. During interviews with pupils I tried hard to ensure they felt comfortable; I believe that an interview should be a 'positive experience' for participants (Kvale, 1996: 36). I attempted to ensure that participants felt able to talk freely, that they felt 'competent' (Rubin and Rubin, 1995: 125) and that they had something useful and valuable to contribute.

Where possible, focus groups and interviews were recorded and transcribed in full; these 'texts' form the basis of much of this study. The texts are limited to what was said without further detail, so observation was important in providing further information about the interactions between pupils and ambassadors. During a few of the one-day events I relied completely on notes taken during conversations with pupils, as there was no opportunity for longer discussions that could be recorded.

I have attempted to maintain an 'ethic of care' (Heath *et al.*, 2009) throughout my research. I felt it was important that participants in activities were aware that I was observing (Gerson and Horowitz, 2003) and that this was explained at the start of each event. I also explained to individuals and groups I had focus groups/conversations with what I was doing and how the research would be used. However, this explanation was brief and pupils were often absorbed in activities or in their interactions with each other, so I may not in reality have managed to acquire fully informed consent. It is arguable that this is ever fully achieved, as the outcomes of research are not fully understood at the start of any project (Heath *et al.*, 2009). However, I have felt my approach to focus groups and interviews avoided some of the traps of exploitation and 'othering' that can occur in interview contexts. While my age, gender, ethnicity and education separates me in many ways from those with whom I was speaking, I was conscious not to exploit this positioning, and was asking the ambassadors and pupils about their experience of the activities and about their relationships with each other without requiring them to give up confidences and personal details. My focus was specifically on their verbal constructions of each other and themselves in relation to each other. The focus of the study on how dominant discourses affect the identities of ambassadors and of pupils rather than presenting participants as 'other', conceptualizes them as part of wider social processes within which identities are formed and within which we all play a part.

I did not feel it ethical or necessary to the study to directly ask pupils and ambassadors for personal details. This was especially important as many of the pupils contributing to focus groups were under 16 and so it would have posed ethical issues if I was to collect pupils' names and personal details without parental consent. I also felt that pupils may not feel comfortable about sharing details about their own backgrounds in group contexts with other young people they did not know well. The study does not use participants' personal details, and participant forenames have been replaced with pseudonyms throughout.

Data analysis

The nature of the data led me to develop a 'toolbox' of approaches to the data analysis. Willig's (2001) suggestion of six specific steps when approaching extracts of data has been helpful, though this has not been followed explicitly. My interest in identifying patterns in participants' accounts means I have fractured the data more than is usually found in analysis of conversations by social psychologists.

Willig identifies as the first stage the need to recognize the 'different ways in which the discursive object', in this case, student ambassadors, 'is constructed in the text'. She explains that it is important to 'highlight all instances of reference' and that this involves looking beyond 'explicit' to find 'implicit' references as well (Willig, 2001: 109). Transcripts and notes from all events were scrutinized and these instances of reference identified.

The second stage involves exploring the differences between constructions of the object. During this phase I was interested in exploring the differences in the ways that participants in the study discursively construct ambassadors and how this related to wider discourses, for example neo-liberal marketized discourses in HEIs.

The third stage is to examine closely the context within which different constructions of the object are made. Willig suggests the following questions:

> What is gained from constructing the object in this particular way at this particular point in the text? What is its function and how does it relate to other constructions produced in the surrounding text?
> Willig, 2001: 110

I found, for instance, that student ambassadors' discursive constructions of an ambassador were in part consciously shaped by what they wanted to say about themselves.

Willig refers to positioning as the fourth stage. I explored the positions available to ambassadors and what subject positions ambassadors took up. Closely linked to the positioning of ambassadors have been the subject positions younger students chose to take up in response.

The penultimate stage focuses on the 'relationship between discourse and practice'. I was interested in the construction and positioning of student ambassadors and how this constrains their relationships with pupils; therefore it has been extremely valuable to explore the impact of these positionings on the practices of being an ambassador in the different contexts considered.

Willig's final stage is one which 'explores the relationship between discourse and subjectivity'. This stage explores the impact of taking up particular subject positions on the subjectivity of those involved. My focus here has been both on the impact on the ambassadors themselves and on how younger students positioned themselves in relation to ambassadors, and on the consequences of this positioning in terms of their 'subjectivity'. It has been vital to this study to look beyond individual conversations with participants to explore the ways in which ambassadors were discursively constructed across the different locations and contexts in which they worked. This required a different approach. Ethnography and discourse analysis have been combined with approaches drawn from grounded theory (Strauss and Corbin, 1990; 1998; Charmaz, 2003). This facilitated a systematic approach to the exploration of emerging patterns in the discourses at different activities.

Strauss and Corbin 'assume an external reality that researchers can discover and record' (Charmaz, 2003: 255). This view is at odds with the poststructuralist and Foucauldian perspective drawn on here, that sees subjectivity as fluid and constantly in creation. Contrary to Strauss and Corbin's ideas that the impact of the researcher can be minimized through 'taking appropriate measures' (ibid.), the research interview is viewed in this study as inevitably part of this dynamic process of identity formation. However, working from the ground up, the bedrock of grounded theory has been vital. The discursive constructions of ambassadors and their work have been traced systematically through the data, and discourses and subject positions identified as a result of that process.

Data gathered during each activity has been categorized and coded into discourses, which have been generated from the discursive constructions (Willig, 2001) of ambassadors and their work in the data itself. This has enabled a genuinely exploratory approach and a fluidity within the data-collecting process: 'Coding helps us to gain a new perspective on our material and to focus on further data collection, and may lead us in unforeseen directions' (Charmaz, 2003: 258).

Central to grounded theory and again vital to this study has been the constant comparative method:

> (a) comparing different people ... (b) comparing data from the same individuals with themselves at different points in time, (c) comparing incident with incident, (d) comparing data with category, and (e) comparing category with other categories.
>
> Charmaz, 2003: 260

This constant comparison has in particular afforded insight into the ways in which pupils respond to ambassadors within the different contexts. While this approach has led to the fracturing of conversations and precluded more intense contextual analysis of extracts of text, such a comparative approach has provided insight into relationships between ambassadors and pupils and into how learning contexts and wider institutional discourses impact on these relationships.

My analysis of the data collected is grouped according to three areas or themes. Chapter 5 focuses on an analysis of discourses of marketing and the related discourses of, first, professionalism and employability at Bankside and, second, deficit and charity at Royal. Chapter 6 presents a series of 'vignettes' about learning contexts, discourses relating to teaching and learning and the related positioning of ambassadors and school pupils. Chapter 7 explores the positioning of ambassadors as 'role models' and the processes of dis/identification between ambassadors and pupils during their work together. The data examined in these three chapters reveal the impact of wider neo-liberal discourses on the outreach work of these ambassadors. Also revealed are the complex processes involved in identifications between pupils and ambassadors, raising issues about assumptions that ambassadors are aspirational role models for school pupils.

Notes

[1] Pseudonyms are used for the two institutions, other organizations, schemes and activities for confidentiality and privacy reasons.

[2] HESA identifies students' socio-economic backgrounds according to the following criteria: 1 Higher managerial and professional occupations, 2 Lower managerial and professional occupations, 3 Intermediate occupations, 4 Small employers and own-account workers, 5 Lower supervisory and technical occupations, 6 Semi-routine occupations, 7 Routine occupations, 8 Never worked and long-term unemployed, 9 Not classified.

Chapter 5

Meanings of marketing

Marketing was a dominant discourse relating to ambassadors among those located within both institutions. This is unsurprising now that HEIs operate as corporate enterprises. Central WP units at both universities were situated alongside recruitment in central administration, and ambassadors were used within this framework as marketing tools for their institutions. There is an inherent tension within the marketing discourses, however, as the Aimhigher funding, which was responsible for a large number of WP activities at the time of the study, was provided to promote progression to HE generally rather than to individual universities.

Practices of employing ambassadors at each institution actually related closely to existing patterns of recruitment. At Bankside, a new university with a local intake, the Aimhigher target group coincides relatively closely with the target group for student recruitment. At Royal, an elite Russell Group institution, to which students are recruited nationally and which requires higher levels of academic attainment for entry, the target group for Aimhigher hardly overlaps at all with the target group for recruitment. Because of this, while some activity was undertaken for Aimhigher with Aimhigher funding, the focus of ambassador work with local pupils funded by Royal itself was solely for G&T pupils.

The discourses surrounding initiatives at Royal and at Bankside differed because of the differences in the positioning of the two institutions. Discourses of professionalism were powerfully evident among staff at Bankside aiming to display the possibilities that study at Bankside offers. Student ambassadors took up and enacted these discourses. In contrast, discourses of charity and deficit circulate at Royal, and students from similar geographic, ethnic and classed backgrounds to those at Bankside appear to take up and enact these discourses in their work with school pupils, dispensing advice and encouragement to the poorer and less-privileged groups that constitute local, minority-ethnic and working-class pupils. The impacts of these different enactments on the subjectivities of ambassadors and school pupils therefore differ between Bankside and Royal.

Promoting university or promoting Bankside and Royal?
Discourses circulating within institutions

The multiple stakeholders involved in the WP work that student ambassadors undertake all discursively constructed student ambassadors as promoters of higher education. There was, however, ambiguity and even confusion in practitioners' accounts about what this meant.

An evident cause of tension was that Aimhigher funding was allocated for promoting progression to university generally, not progression to particular institutions. When working with pupils, such a distinction between promoting progression and promoting particular universities is evidently difficult to make. It is somewhat contradictory that the student ambassadors are marketing university but not particular universities. Indeed, this contradiction was embedded in an Aimhigher coordinator's explanation of what Aimhigher were funded to do; while explaining that 'marketing isn't on the list' of activities funded by Aimhigher, the coordinator went on to say that Aimhigher is 'marketing the idea of progression'.

The impracticality of distinguishing between marketing progression generally and marketing progression via particular institutions was clearly illustrated by the Aimhigher coordinator's account of how ambassadors were asked to work with pupils:

> I think one of the strengths of being an ambassador is that you are a witness and therefore to say to somebody, 'don't say which university you've come from', I think actually isn't what the person who goes up and talks to you wants to know. I'm not saying that an ambassador should go, 'I'm from Royal', but that actually if somebody says, 'where are you from, what are you doing', they answer … I mean there is a clear danger that a university will hijack a nice pot of money, a nice extra set of staff; in reality I would rather the student ambassador answered the questions that whoever is talking to them wants answered, than we were very, very strict about, 'don't say where you're from'.
>
> Aimhigher coordinator

The difficulty of ensuring that student ambassadors 'market' progression routes rather than individual institutions was compounded by the fact that many worked on recruitment activities for their own universities as well as on WP activities. The only tool available to WP teams and Aimhigher staff to ensure that student ambassadors differentiated between marketing

Meanings of marketing

university generally and marketing specific institutions, was a different coloured T-shirt:

> In the training, what ambassadors are often told is, which T-shirt are you wearing; if it's an Aimhigher one, you're talking about progression; if it's a Bankside one you're at a Bankside open day and you're talking about courses at Bankside and you need to know about courses at Bankside ... Aimhigher is different ... because it's offering knowledge of a pathway, not knowledge of a specific ... .
>
> Aimhigher coordinator

The physical positioning of the WP units at Royal and at Bankside was also revealing. At both institutions, the WP teams were managed by marketing directors and both teams were physically located within the marketing department. One of the WP coordinators at Royal described team meetings as 'very marketing focused'. The WP manager at Royal complained that 'internally we are always told to do recruitment' and at the end of her conversation with me actually posed the question 'are we recruiting?' and appeared genuinely unclear about her position.

It was evident from WP staff responses at both universities that more attempts had been made in the past to distinguish between the marketing work of student ambassadors and their WP work. However, the impracticality of maintaining this separation and internal pressure from management within the university had ensured that distinctions had blurred and faded over time:

> Initially I was very clear I was working for Aimhigher. I would say to student ambassadors, 'don't talk about Bankside'. I've now moved to a position where I think they should talk about what Bankside offer ... they are proud of their own university and particularly if they are subject-specific students on a vocational project, they should be able to talk about their course and what they do, but I wouldn't tell them to promote Bankside.
>
> Outreach manager, Bankside

The WP unit at Bankside had previously been located in a different office, separate from the marketing department. At Royal, WP ambassadors and ambassadors used for marketing the university had, at one stage, been employed separately. The WP coordinator there suggested that the amalgamation of schemes was 'driven' by student recruitment:

> Two years ago there were murmurings of 'oh, it should just be one scheme, there should be one entry point overall into the Student Ambassador programme' … That was driven by, at the time, two people, actually, who were in Student Recruitment.
>
> WP coordinator, Royal

WP units were in a weak position within both universities, but especially at Royal. They were largely externally funded by Aimhigher, were not part of any academic discipline and were situated somewhat uncomfortably in a no-man's land alongside administration departments. While at Bankside the target groups for Aimhigher matched the young people that Bankside would target in recruitment activities, this was not the case for Royal. These young people were of no interest to the recruitment department at Royal:

> The problem is the type of student, at Bankside they recruit locally from London so it works together, but Royal's recruitment is national. … recruitment teams from Royal would never go to these schools.
>
> WP manager, Royal

Royal funded the WP unit to work with some local schools, but only with G&T pupils, and even then there was only a 'small pot of money' available.

Given the placement of WP units within both universities, it is unsurprising that in both cases these units were subsumed by the larger, more dominant and powerfully positioned marketing teams:

> Slowly over time the internal pressures mount. There is a benchmark [relating to student numbers] and you are asked what you are doing to meet it? You get – it's lovely you're doing charity work but …
>
> WP manager, Royal

The intertwining of widening participation and recruitment was notable during the meeting at Bankside involving a range of stakeholders who worked with student ambassadors, including the head of recruitment at the university, the outreach manager and other members of the WP team, as well as borough coordinators from the Aimhigher partnership. The focus was the recruitment of student ambassadors. Marketing discourses ran throughout the discussion and were taken up by all stakeholders.

The outreach manager spoke about the need to 'create a group' who 'embody the most successful graduates of Bankside'. This manager also talked about the possibility of having 'super student employees' who could 'represent the university at high profile events', describing 'a core of experienced, well trained students who we can rely on to train others, represent the university at high profile events and to the media and lead on project work'.

These discursive constructions of ambassadors draw on discourses relating to marketing, made explicit by the head of recruitment at the university, who went on to discuss how the ambassadors needed to be 'professional' and 'corporate' since they are 'representing the institution they are working for'.

This focus was discussed as important to the training of student ambassadors, and the need for training to focus on the institution to instill a 'sense of pride' was agreed on by all present:

> They should have an introduction to the institution to give them a sense of pride.
>
> Schools and colleges outreach manager, Bankside

> They should have low-level corporate induction – then build it up in later training.
>
> Borough coordinator, Aimhigher

As Ball (1994; 2003) suggests, in these contexts it seemed that individuals, both within institutions and outside, took up and enacted government policy. Dominant personalities at Royal were identified as key in developing the unified student ambassador scheme incorporating WP ambassadors and those involved in recruitment. It is important to relate these micro-level ways in which government policy is practised in institutions with the macro level: the wider discourses of the marketplace circulating in HEIs across the UK. These extend beyond individual institutions. The ideology of the marketplace operates as a regime of truth dominant in the contemporary UK HE system.

Royal: marketing discourses circulating at events

Marketing discourses were present in ambassadors' and pupils' accounts of their work together. The confusion expressed by the WP coordinators about whether ambassadors were promoting Royal or university more generally was also present among student ambassadors. While there was some suggestion from ambassadors' accounts that they were promoting university generally, it was clear that they perceived promoting Royal to be part of their work in all contexts, including WP activities. The pupils' accounts indicated they were aware that ambassadors were promoting Royal and this was often effective, with pupils frequently expressing an interest in going to Royal in the future.

During the focus group held with ambassadors at the G&T summer school, ambassadors explained that there were different training days for the different types of work that they undertook. However, for some, little distinction was made regarding the type of activity that they were involved with:

> Than [student ambassador]: I applied, I did some Open Days when I was a student ambassador, like last year and then I filled in an application and became a student ambassador and they put me on the widening participation mailing list as well so, to be honest, I don't know the difference between them 'cause all I do is, I get emails from widening participation, Royal – I just say, 'yes', 'no'; they say, 'training days', I say, 'can I do it' – if I can do it I do it – if I don't … so I wouldn't really know what was what to be honest.
>
> G&T summer school

Student ambassadors' accounts during the medical days at Royal also illustrated this ambiguity. The ambassadors' discursive constructions of their work in these contexts did not specifically focus on promoting university. Their accounts were, however, full of references to wanting to 'help' pupils; one ambassador explained that part of what drew her to working as an ambassador was that 'you can influence' pupils:

> Chanelle: Yeah, I enjoyed working with kids before I came to uni … I was like, that's something I could really enjoy especially when you can influence them to have a positive impact in their life.
>
> Medical day

Encouraging pupils to progress to university was evidently part of the 'positive impact' that this ambassador hoped she would have. School pupils' accounts echoed this, with one pupil explaining that the focus of ambassadors' questions was about going to university and another explaining that ambassadors 'try to encourage you'.

Field notes taken during a tour of the campus, however, suggest that ambassadors were more specifically promoting Royal. One light-hearted exchange focused on the benefits of having a Burger King on campus, how good the university is and the 'good experience' of being there:

> Candice: It's the only uni in the UK with a Burger King on campus.
>
> Abiola: I want to go to this university – there's a Burger King here.

Meanings of marketing

Candice: Yeah this uni is top. University is a lot of work but I think it's a really good experience and I'd encourage as many people as I can to have that experience.

Medical day

When asked about the ambassadors' work and what they were there to do, one pupil commented that they were providing information about university, and Royal in particular:

Ton: To help us learn about medicine and all that and how university is really like and also, 'cause like one of the ladies told me this university here is mostly about medicine and there's another one in Castle that specializes in all those other like English, law and others.

Medical day

Another pupil explained that going to Royal, a 'good university', would be beneficial in the longer term:

Anwa: I think it's a kind of good university 'cause once you're there you may feel bored sometimes but at the end you get the job that you want so it's kind of worth it.

Medical day

This pupil appeared to have assimilated the view expressed in very similar terms by an ambassador during the day as his own. The marketing strategies employed by the ambassadors therefore could be seen to be effective; this G&T school pupil appeared to be imagining himself at Royal and to be positioning himself as a future university student.

The active promotion of Royal was more explicitly evident during the G&T summer school. Ambassadors' accounts demonstrated that they were aware they were supposed to have different functions at different events, such as working for the recruitment department and promoting the university, *and* also as an ambassador involved in WP work. However, they discussed how their behaviour did not change. Munira and Casey explained that as student ambassadors they were 'representing Royal' at all events:

Munira: Because you're trained for the one job, not trained to go, 'well, if this is a Royal event this is how you behave and if it's an Aimhigher event this is how you behave'.

Casey: 'Cause we're still representing Royal in one way or another.

G&T summer school

Martin's account of the presentation given by ambassadors on the first day of the summer school certainly indicates that ambassadors were presenting a very positive view of life at the university:

> Martin: I mean they taught us a lot about, like, the courses that we were mostly going to be taking and medicine, biomedical sciences and things like physics … the campus and what kind of things you get and the different clubs they have here – like societies and the student union and all of this different stuff; it's been quite interesting … student life, bars.
>
> G&T summer school

Further conversations with school pupils revealed that the ambassadors were explicitly promoting Royal and encouraging pupils to apply:

> Imogen: They gave us a talk okay and the campus – sort of showed us round the places, gave us tips.
>
> Imogen: We talked about Queens as well. Quite a lot of rivalry between Queens and Royal and they were talking about how Royal is more relaxed and there's more of a social life and Queens they all tend to be social outcasts … so that's quite funny.
>
> G&T summer school

This direct approach by the ambassadors, in the short term, appeared to have been very effective. One pupil explained that the ambassadors had 'sold' Royal to her:

> Vanessa: I didn't really know about any universities until I came to this one. I like this one; it is really good.
>
> Lola: Yeah it is.
>
> Vanessa: They've sold it to us …
>
> G&T summer school

Another pupil explained how the ambassadors had 'won us over'. The pupils described how 'friendly' the university is and how 'relaxed and casual'. It is reasonable to conclude that the ambassadors were largely responsible for these perceptions of life at Royal:

> Clare: Do you know which university you're interested in going to?
>
> Lola: Yeah – they won us over.

> Martin: Maybe – we'll see – see where I get into first ... but it's been really nice here; it's a really friendly university – really.
>
> Clare: Yeah, what is it about it that's won you over?
>
> Lola: It's very relaxed and casual.
>
> G&T summer school

The pupils also talked explicitly about how the ambassadors marketed the university to them, referring to how the ambassadors 'talked it up quite well':

> Kate: And the ambassadors, they talked it up quite well – the social life as well as getting the good grades.
>
> Martin: Everyone seems to smile here except the librarian.
>
> G&T summer school

The focus the ambassadors placed on the social life at university seemed to appeal to pupils in the focus group. These pupils also appeared to have strong identities as learners, and the ambassadors' emphasis on the balance they achieve between work and social life was clearly important.

The pupils' awareness of the way ambassadors market the university was explicitly referred to. Pupils were amused by ambassadors' protestations that they are not 'selling' Royal; they were aware but open to these marketing advances from ambassadors:

> Vanessa: ... they just made it seem really good here.
>
> Lola: And they kept saying, we're not trying to sell it to you.
>
> Vanessa: Yeah, but they made a pretty good job of selling it [laughs]. Yeah, just emphasizing how they can still get the grades and still become what they want while also having a social life and all the clubs we can join.
>
> G&T summer school

It is also worth noting that two of the ambassadors themselves had worked with ambassadors at Royal when they were at school and that this attendance at events had contributed to their decisions to apply:

> Munira: I did one of the Aimhigher events when I was at sixth form and that's how I got introduced to Royal.
>
> Alicia: I actually came to the summer school – the medical one. I loved it so much I came back.
>
> G&T summer school

The ambassadors were consciously marketing their university to this group of pupils and were doing so effectively; pupils were positioned as consumers and appeared to be willing participants in this marketing process. However, given that the G&T pupils I spoke to during the summer school appeared to be from middle-class backgrounds, the ambassadors' work with them is unlikely to challenge existing patterns of exclusion of poor and working-class young people at Royal.

Bankside: marketing discourses circulating at events

At Bankside, student ambassadors were also promoting university generally and Bankside in particular, and pupils' accounts again indicated that they were aware of these marketing strategies and were open to them. While pupils were positively oriented to the ambassadors and interested in finding out about Bankside, there was less indication from pupils than at Royal that they aspired to progress to Bankside. However, in the longer term, the positive experiences pupils had at Bankside could encourage progression to the university. There were, however, instances where pupils appeared less positively oriented to these promotional strategies.

During the engineering camp, one student ambassador explained that she used 'any opportunity to tell them [school pupils] about university'. Her use of the phrase 'we're selling that' draws explicitly from discourses of sales and marketing:

> Gill: Any opportunity we tell them about uni – in a subtle way like 'you need a degree for everything!' – we're selling that.
>
> Engineering camp

At the summer school, ambassadors also referred to their responsibility to promote university and, as at Royal, pupils shared ambassadors' awareness of this marketing focus. One pupil explained, uncritically, that the ambassadors were there to 'try to make' university 'seem ok'.

Again, this reflects an understanding on the part of the pupils that ambassadors are there as marketing tools to encourage progression. Being positioned as consumers in this way, however, seemed so familiar to pupils that they viewed it as hardly noteworthy. That the HE system is a marketplace and that they are consumers within it is a regime of truth that makes this a seemingly natural positioning for them.

Despite awareness of their positioning as consumers, evidence from observations and interview data revealed pupils' interest in finding out about university from ambassadors. Their interest in hearing information from ambassadors about university positions pupils as willing participants in this marketing process. The ambassadors described providing pupils with detailed information about how they work at university. Qadira relayed conversations she had with pupils about lectures and other modes of teaching at Bankside:

> Qadira: They do ask a lot of questions about lectures. Remember when they were asking about lectures because of that lecture that we had they were like – oh wow that's really long and boring. We had to explain to them that that's how some lectures are but once you get to a certain stage where you actually want to know the information you don't find it boring anymore and then we try and explain to them the differences between like seminars, lab sessions, lectures as well so they wanted to know basically like what university is like – a lot of questions about that.
>
> Bankside summer school

Other exchanges described by ambassadors included information about where students live:

> Adam: They want to see how much they can get comfortable around uni as well. I can remember yesterday one was asking me 'can I stay at home or do you come from home or do you stay in halls of residence?' Lots of questions like that … most of them want to stay away from home. They want to go in the halls of residence.
>
> Bankside summer school

Qadira recounted pupils' repeated questions about what she studies and how university is different to school:

> Qadira: For me it's the same actually. They always ask the same type of questions whether we're in the afternoon activity – it's even more funny because in the afternoon activities we've got students from different strands and they're always – 'oh, what strand are you on' – like, 'what do you do' – and I'll start explaining again and they'll be asking – 'do you like this uni', 'how's the uni like', 'is it like school', 'do you have to do this', 'so you have to do that?' It's the same questions.
>
> Adam: Pretty much the same. It's usually the same questions – it's like – 'what do you do?'
>
> Bankside summer school

As well as general questions about university life, ambassadors described being asked specific questions about Bankside: 'do you like this university?' and 'how's this university?'

Clare Gartland

Student ambassadors were positioned as promoters of university generally and of Bankside in particular, providing pupils with specific information about Bankside. Unlike Royal, however, pupils did not talk about wanting to go to Bankside, despite their interest in finding out about it. School pupils are likely to be aware of the relative perceived value of different universities and the degrees that they offer. However, the data indicate that these pupils' interest in Bankside and their experience of being there may contribute to their viewing it as a 'comfortable' option (Reay *et al.*, 2005). It could be that this experience of being at Bankside encourages them to apply. These pupils all discussed plans to progress to HE that were established before embarking on the summer school. They were already on a trajectory to HE. As with the G&T pupils at the Royal summer school, their experience of being at Bankside may just serve to reinforce existing patterns of HE participation: 'while more working-class and ethnic-minority students are entering university, they are generally entering different universities to their white middle-class counterparts' (Reay *et al.*, 2005: 162).

However, the ambassadors' marketing practices were not always effective. One of the ambassadors at the engineering camp made an important distinction in the attitudes of pupils attending the camp. Unlike the summer schools at Bankside and Royal, where pupils discussed having already decided to go on to university, pupils attending the engineering camp had not all decided to go. Gill commented that it was only pupils already set on going to university who were keen to engage in discussion with ambassadors:

> Gill: The kids that are keen, they do ask questions about uni and engineering – some kids know that uni is the thing for them so they are asking about it, like 'What do you do in the evening? Do you have to work really hard?'
>
> Engineering camp

A few pupils expressed some resistance to the marketing approaches taken by ambassadors. One was even irritated by ambassadors telling her about 'the stuff you can gain' from university:

> Fabienne: The supervisors didn't really help me. It's three years til I go to university – I don't really want to know – I'm not bothered. I already know what you have to do – I knew the stuff that you can do – stuff you can gain from it.
>
> Engineering camp

This poses questions about why these pupils responded differently to the marketing advances of ambassadors to pupils on the summer school. That they are Year 9 instead of Year 10 or 11 may have contributed, though there were other contexts where younger pupils responded positively. While it may be that some pupils on the engineering camp had less-established learner identities, Fabienne herself was keen to attend university and wanted to study engineering, so she had a positive learner identity and shared subject interests with ambassadors. However, I argue that the context was significant and that ambassadors were positioned differently during the engineering camp (Gartland, 2014). The influence of different learning contexts is considered in the next chapter.

Selling subjects

The student ambassador schemes at both universities were not managed and organized by the WP units alone; there were other stakeholders involved in the administration and organization of ambassador work, especially the AEP and MAS. These stakeholders were focused on WP in STEM subjects, particularly engineering and medicine.

Bankside: Selling engineering

At the time of my study, the AEP was running at Bankside. Student ambassadors were employed for this project to work with pupils both in schools, at the university and at other venues. Discourses relating to promoting engineering as a subject were strongly represented in the accounts of organizers and ambassadors. Ambassadors promoted messages about engineering, including the job opportunities available and the opportunities for creativity and fun that engineering provides. These marketing strategies had some impact on school pupils, although the age of pupils and their established learner identities and orientations to subject areas influenced the efficacy of such strategies.

Student ambassadors studying engineering or a related STEM subject were given further training by AEP fieldworkers after training by the WP unit. This involved training about promoting STEM subjects to pupils and promoting 'engineering messages' with a view to raising awareness among school pupils about engineering careers and encouraging progression into either engineering degree courses or into STEM more generally. The opening slide at the start of the training session identified that the student ambassadors were there to 'promote STEM careers and study' and to 'promote engineering messages'.

One AEP fieldworker described succinctly that ambassador work should 'promote engineering messages, challenge the stereotype of engineer as mechanic and make university seem accessible'. Student ambassadors

69

at all events were aware that they were expected to promote engineering. Indeed, student ambassadors' discursive constructions of their work at all events related to promoting engineering and promoting engineering courses at university: 'We were told in training – encourage science, encourage engineering, encourage university' (Train Tracks). One student ambassador explicitly drew on this marketing discourse to describe her understanding of the work of ambassadors subsequent to the AEP training as 'getting the students to think about engineering subjects like science and maths ... promote it in a good way'.

The ambassadors' accounts of approaches taken with pupils closely reflected the engineering messages they were introduced to during training. One message was that engineering can be fun, and this was referred to by ambassadors at various events:

> Akua: The job is motivating them – showing them what it's going to be like – it's not different from any other course. Some of them are saying it's hard – it's showing them it can be fun – find a solution not just lots of reading ... it's not as difficult as people make it out to be.
>
> Engineering camp

Ambassadors were also involved in promoting engineering through discussing possible available salaries and different types of engineering jobs. During the engineering camp, one ambassador described how she stressed the benefits of being an engineer in terms of job opportunities. She seemed to be consciously attempting to market engineering to pupils as a glamorous international occupation:

> Akua: I said, 'if you do engineering you have a profession – you can work anywhere – whereas something like drama is not the same in Ghana as it is here ... engineering you can still follow your career anywhere' ... I said, 'you can work in Kuwait, San Francisco – you can go anywhere'.
>
> Engineering camp

This ambassador also described how she had provided information about engineering careers that related to pupils' own interests:

> Akua: He said, 'I like designing, drawing ... I liked working on the project', and I told him about civil engineering, and he said, 'can you work anywhere in the world?' and I said, 'yeah' and he said, 'on small or big projects?' and I said, 'small and big projects' ... I

can throw more light on it – he doesn't like planes – he wants to go into construction. So now if he fills in his UCAS form he knows what to put on it.

Engineering camp

Another strategy employed by the ambassadors I observed was to try to raise pupils' awareness of jobs in engineering by talking about the roles they were taking on during the practical activities they were engaged in. I overheard one ambassador explaining to a pupil that his role in the practical group task was 'to be technical manager' (careers afternoon). At the careers afternoon, possibly partly because of the careers focus of the session, pupils were open to messages about pay and types of jobs and commented about what they had learnt:

> Isima: I didn't know you got paid that – it's quite a lot for that.
>
> Amber: I've learnt about engineering jobs – there's lots of different types.

Careers afternoon

Another engineering message that ambassadors were actively promoting was that engineering is a creative subject. This message was relayed to school pupils in a conscious promotional strategy during the careers afternoon. During a short dome-making exercise that ambassadors facilitated, one commented on the need for a design focus during his explanation of the task: 'Engineering is not just about building; you've got to make it look nice.'

This message was absorbed by some pupils; one girl commented to me that designing was something she had enjoyed about the work, though she was somewhat ambivalent about whether this was required of her in the task: 'I like designing – even though that's not their aim – the designs are quite pretty' (careers afternoon). Another ambassador identified this focus on creativity as motivating for pupils:

> Ed: We were talking about their future and jobs – what they want to have. One said if engineering is about being creative they'd be interested … some of them were really like – yes this is what I want to do.

Careers afternoon

One female ambassador relayed a conversation she had had with a pupil about her car design, suggesting this pupil's engagement with the creativity involved in the task:

> Kelly: I said, 'if this was your car in the garage what colour would it be – I work for VW how would you sell it to me?' And she said: 'it's purple and it's got hearts on so it's for girls'.
>
> Careers afternoon

While this emphasis on creativity may be an effective marketing strategy, it is worth considering the actual activities that younger students were undertaking during these discussions. The relationship between the discourse and practice is important here. In encouraging pupils to link creativity and engineering during small-scale accessible tasks, it may be that ambassadors are failing to provide pupils with accurate information about engineering or are even misleading them. The girls' enjoyment of the colour and decorations on their car does not relate very closely to the more technical aspects of design required on engineering courses. Indeed, rather than challenging gendered identities in relation to engineering, this positioning of pupils may serve to reinforce gendered identities in relation to the technical aspects of engineering.

From the accounts of both ambassadors and school pupils it appeared that pupils were more interested in the ambassadors' own experiences than in being told more generally about engineering. During Train Tracks, an ambassador's account of her interactions with pupils suggested that the pupils initiated this focus:

> Rachel: They were asking about my degree and university and where it leads.
>
> Train Tracks

This was also reflected in the accounts of the ambassadors and pupils at the Bankside summer school. Pupils were able to identify the subjects ambassadors working with them were studying:

> Michael: One was doing explosive stuff … .
>
> Dina: He's doing fire and explosives.
>
> Joe: Science and engineering … it's like forensics or something. Most of them are doing engineering.
>
> Sas: Yeah and one of them, Dow, was doing chemical engineering.
>
> Sarah: Someone was doing architecture.
>
> Michael: They are all doing something related to engineering.
>
> Bankside summer school

Meanings of marketing

Qadira explained how she had talked to pupils about studying at university; the course she described to them was her own engineering course:

> Qadira: They do ask a lot of questions about lectures. ... we try and explain to them the differences between like seminars, lab sessions, lectures.
>
> Bankside summer scool

Ambassadors described frequently being asked about their course:

> Adam: It's usually the same questions. It's like: 'what do you do?'
>
> Qadira: A lot of the girls say: 'oh, so what engineering do you do?'
>
> Bankside summer school

During Train Tracks, conversations about the ambassadors' own experience, when combined with the day's activities, were effective in raising these pupils' awareness of engineering. The pupils had absorbed some key engineering messages:

> Sarah: [Engineering] ... there are some difficult things to consider ... I wouldn't say it's an easy subject – it's something where you'd need to use your initiative – you need to put other people into what would go wrong and what would go right.
>
> Ayisha: You need to plan it all out exactly ... it's about team work ... although it's complicated you are able to work it out in small stages so you will eventually get there ...
>
> Sarah: Engineering before – I probably thought it was to do with mechanics and how things work but now I have a broader view of it – except I can't say it in words.
>
> Meena: You know chemical engineering involves science and D&T? ... I'm just interested ...
>
> Train Tracks

The combination of the ambassador's input and the day's activities was successful in promoting positive orientations to engineering careers among this group of girls:

> Sarah: Engineering is another word for making things.
>
> Ayisha: I want to become an engineer now.
>
> Meena: Only if it involves science.

> Aiysha: Yes for the rest of my life.
>
> Train Tracks

However, this positive response to the ambassadors' marketing of their subject area was not guaranteed. On the engineering camp, despite the enthusiastic accounts of various ambassadors about how they were promoting engineering jobs, corresponding enthusiasm was rarely found in the accounts of pupils who, in many instances, appeared unaware that the ambassadors were engineering students. Many pupils did not hear these messages. In this context, the student ambassadors were clearly not viewed by school pupils as 'hot sources' of useful information (Ball and Vincent, 1998; Ball *et al.*, 2000; Archer *et al.*, 2003).

The data indicate that other issues relating to the success of this marketing of subjects was the age of pupils, their learning identities and their established identities in relation to particular subject areas. While pupils who attended the summer school were interested in the student ambassadors and finding out about their life at university, there was little evidence that this interaction changed pupils' minds in terms of their own subject choices.

One pupil talked with admiration about the fact that the ambassadors study engineering:

> Michael: They study engineering here so we know they are much smarter than us.
>
> Bankside summer school

But some pupils' accounts reveal that they perceive engineering and robotics in particular to be difficult – 'hard work' and 'complicated'. A particular lecture heard during the summer school and their attempts at programming a robot had contributed to this perspective:

> Sarah: The lecture was – that killed me. That killed me.
>
> Dina: I liked the videos.
>
> Michael: It was because he didn't put any enthusiasm into it, you know. He didn't have any charisma at all.
>
> Clare: Does it appeal to you, making robots? How does the whole thing make you feel?
>
> Michael: Seeing how hard it was to programme it I don't think I really want to try.

> Dina: Yeah, he was saying for the arm it took them like – what was it – five years to build it and that's just the arm.
>
> Sarah: That's long.
>
> Sas: Personally, I couldn't be bothered with all that work. My view of life is to grow up and make my money.
>
> Sarah: And the thing is, if you make a mistake, you won't know what you've done 'cause it's so complicated, and you'd have to get it right to make sure it doesn't malfunction and stuff. It's just, like, so complicated.
>
> Bankside summer school

The pupils came to the summer school with ideas about subjects they wanted to do after their GCSEs. Two of the pupils in the focus group were interested in engineering-related subjects at university, and the summer school, as with Train Tracks, appeared to confirm and reinforce that interest. The others, however, had not selected engineering as their first choice for the summer school and had interests elsewhere. The summer school had not switched their focus to engineering. As research suggests (Ormerod and Duckworth, 1975; Tai *et al.*, 2006; Office for Public Management, 2006; Archer *et al.*, 2010), pupils' identities relating to science subjects are already well established by this stage. Indeed, the encounter with the engineering lecturer, though perhaps mitigated by working with the ambassadors, may actually have served to reinforce preconceived ideas about scientists generally and engineers in particular – he 'didn't have any charisma at all'.

Selling medicine at Royal
The Medical Access Scheme (MAS)

At Royal, the student ambassadors on the MAS were separate from ambassadors employed centrally. The purpose of these student ambassadors appeared to be direct recruitment for the MAS. These ambassadors' discursive constructions of their work focused on encouraging students to apply for medicine and particularly promoting the MAS:

> Simi: There was a time we had a session, there was a girl from my school and I spoke to her and she was like, 'so you do medicine', and I was like, 'yeah', so I started explaining to her the programme that we're on … . She goes to the same school, same sixth form and they're still not advertising the programme. In the end she applied and I think she got in. She's waiting for her grades but it's showing that by us doing this we are able to go back to our schools, speak

> to them about the programme because for some reason the schools are not too happy to advertise it. They think it's like a cheat into medicine; however we still do the same exams; we do the same studies, everything, so this programme allows us to do that, to kind of continue that.
>
> Medical day

Another ambassador talked proudly about providing pupils with information about the programme that resulted in their application. This was again viewed as a success:

> Candice: Two girls I spoke to yesterday, they were in sixth form already; they were quite interested in the course that we're on, the MAS because they understood how they worked themselves and they realized that with the five-year course – they were quite scared of the five-year course because they thought they would just be pushed into something, and then they wouldn't be able to settle in and understand as much as they would want to understand, and they liked the fact that with the MAS you get a little bit more support and they understood the fact that they would get that support, and yeah – so I was just advising them or telling them about the MAS and they realized through that, that okay, this is something that they would like to look further into and then they started saying, 'okay, I think this would be a serious application', actually that 'I am going away to fill my application form in' – that 'this would be a subject that …' – yeah.
>
> Medical day

In terms of action orientation, it is clear that, for these two student ambassadors, pupils understanding the benefits of their course and seeing it as a valid way forward and not a 'cheat into medicine' is important. The importance of this is perhaps rooted in these ambassadors' insecurities about their own positioning on a course that marks them as different to other medical students on the standard medical degree programme at Royal.

SELLING MEDICINE AT THE SUMMER SCHOOL … 'THEY HAVEN'T TOLD US ANY OF THE NEGATIVE STUFF'

For the pupils I spoke to who were interested in studying medicine the G&T summer school had been highly effective in promoting the course at Royal. Pupils talked about how the ambassadors had provided advice about the challenges of studying medicine:

Lola: A lot of things I have been to like this they've been very, 'medical school is very hard, you're not going to have time to go out and be with your friends, you'll be in every night', and they are, like, 'no'.

Martin: Whereas these guys are quoting the bar times to us.

Lola: They are all real about it, I think ...

Martin: They just try to make it, like ... whereas before you'd be like, wow, it's going to be really hard, and they are, like, 'just take it slowly, do your work as it comes and you have to work hard but it will be fun'. They've highlighted the positives – they haven't really told us any of the negative stuff.

G&T summer school

The pupils said they were reassured by the ambassadors' accounts. Lola had previously been told that studying medicine 'is very hard' and there would be no time 'to go out' or 'be with your friends', whereas the ambassadors had countered this view. Martin's account reflects this, describing how the ambassadors had been 'quoting bar times' and explaining that the work is manageable. However, together with this acceptance of the ambassadors' accounts of managing study alongside a social life, Martin's discursive constructions reveal that he is aware of his own positioning as a 'consumer' of HE and that ambassadors are marketing their course. He references this implicitly when he describes how 'they try to make it ...' and more explicitly when he comments 'they haven't told us any of the negative stuff'.

Bankside: professionalism, customer care and employability

While marketing discourses were prominent in the talk circulating among staff, students and other practitioners at both institutions, there were also real differences in how the work of student ambassadors was discursively constructed. Discourses relating to marketing were taken up and enacted differently by employees at each institution. At Bankside, a dominant related discourse was the professionalism of student ambassadors. The dominance of this discourse resulted in the professionalization of ambassador recruitment and employment and impacted on how ambassadors discursively constructed their work. The focus on professionalism also relates to a wider discourse of employability that circulates particularly within new universities.

At Royal, the interview process for student ambassadors was held in a seminar room and was not actually a selection process. Students were accepted as ambassadors on the basis of their written applications. The interview merely confirmed their appointment and was a short and informal process. During the interview, interviewees were told that the money they earn could not be relied upon as an income. This approach contrasted to the professionalization of ambassadors at Bankside. This focus on the professionalism of ambassadors related closely to marketing discourses, with pupils and teachers in schools being constructed as customers. This was reiterated on a number of occasions at the meeting at Bankside: 'We need to be reinforcing messages about customer care and health and safety' (Head of recruitment). This discourse of professionalism was taken up by all stakeholders and not just by the head of recruitment. The outreach manager identified the 'current lack of customer focus and professionalism'. One of the borough coordinators suggested the possibility of student ambassadors gaining 'professionally delivered customer care qualifications'. (The practice of awarding such qualifications was cited as already being practised by Aimhigher in another region.)

New contracts had recently been developed for ambassadors at Bankside that both increased their pay and the commitment required from them. These contracts were identified at the meeting as 'an opportunity to professionalize the role of ambassadors'. There was also discussion of offering ambassadors a guaranteed income to further professionalize the role, the need for a 'more rigorous recruitment process' and continuing professional development (CPD):

> We need to raise the bar … a more rigorous recruitment process, professional training to raise their expectations about what they will be doing and CPD, self-assessment and a probationary period so that they make more effort in passing and we should get regular feedback from customers.
>
> Outreach manager, Bankside

The emphasis on professionalizing the ambassadors and their work and focusing on 'the customer' was evident here and in the ensuing recruitment process and training observed. Ambassadors were recruited through assessment centres; the process took all day and was formal and rigorous, clearly and consciously emulating practices in industry. The assessment took place in plush offices with comfortable padded office chairs, not in the lecture theatres or seminar rooms more normally associated with students.

These discourses also appeared in ambassadors' accounts of their work with school pupils. During the STEM day at Bankside, two female ambassadors talked about their work as similar to customer service. This was described as being an important aspect of the work ambassadors do for the UK recruitment unit at the university, and particular tasks were seen as being similar to those that would be expected of a 'customer services executive':

> Freya: Working on course enquiries is most like being a customer services executive. You organize, give tours, book rooms.
>
> STEM day

This discursive construction of their work as 'customer services executives' was also applied to the WP activity they were doing within the AEP at the STEM day:

> Wendy: Working for the AEP is like customer services. You treat the children with respect so that they're nice back to you.
>
> STEM day

Here, Wendy was talking specifically about the importance of treating customers generally, and school pupils in particular, with respect. Later on in the conversation, Wendy stressed the need to be 'approachable' as ambassadors, a construction again suggestive of customer services.

At the summer school too, there were discursive constructions of ambassadors that could be related to customer services. These ambassadors also worked for the UK recruitment unit dealing with enquiries, working on the front desk and showing visitors and prospective students around. Both ambassadors during one conversation referred to how they had been 'showing [pupils] around'. The pupils themselves talked about how the ambassadors help them and how 'they are very, very helpful'. Both ambassadors made several references to their work 'assisting' school pupils, a term often associated with retail. It is possible to suggest that ambassadors and pupils drew on discourses related to customer service, as there was a shared understanding that ambassadors were being paid specifically to help pupils, again positioning the pupils as consumers of HE and the ambassadors as marketing professionals.

It is significant that this discourse of professionalism was so much more evident at Bankside than at Royal. The explanation for this difference can be found in the different positioning of the institutions within the HE marketplace. These discursive constructions relate to employability discourses that task universities, and particularly new universities, with developing students' skills to increase their employability as graduates (Morley, 2001; Brown and Hesketh, 2004; Moreau and Leathwood, 2006; Hey and Leathwood, 2009). The designation of non-traditional students as having particular and special needs (Burke, 2006; 2012) and their being perceived as less employable than their counterparts at elite institutions has contributed to this focus on employability within new universities. Discourses relating to the employability of students are so well established at Bankside and within the public sphere more generally, that they are taken up as natural by stakeholders involved in organizing the work of student ambassadors. These discourses are then also taken up and enacted in the practices of student ambassadors. Ambassadors are involved in practices that resonate with those found in retail. Being positioned as consumers is extremely familiar for school age pupils in our society, but this positioning could undermine the relationships that these young people are able to develop in these contexts. While familiar, being positioned too clearly as a consumer may impact on the subjectivity of pupils by eroding rather than building up the trust that appears to be so important to young people if they are to access the information on offer. Student ambassadors' positioning as 'hot sources' of information who can be trusted by younger pupils may be threatened by too overt a positioning of student ambassadors as marketing professionals and of pupils as consumers.

Discourses of charity and deficit at Royal

Members of the WP unit described the way the marketing department at Royal viewed their Aimhigher work as 'charity work'. Discourses of charity and deficit also circulated among organizers and ambassadors, together with discourses of individualization and the power of individual students to overcome difficult circumstances. These discourses were particularly dominant during medical days, where the MAS student ambassadors identified closely with the pupils they were working with.

During the training for WP ambassadors provided at Royal by local borough coordinators, discourses of deficit were clearly present. This was indirectly referenced through comments describing how complicated the HE system is and through references to the transformational nature of education. There were also explicit references to the sorts of pupils targeted by the term widening participation. These pupils were described as deprived and as lacking aspiration. Pupils are identified as problematic through these discourses (Burke, 2002, 2006, 2012; Archer and Yamashita, 2003; Yorke and Thomas, 2003; Watts and Bridges, 2004; Bridges, 2005). Students undergoing training also drew on these discourses of deficit, talking of their own underprivileged area or of having come from a 'bog standard comp':

> Jessica: I'm a bog standard state student – I got mediocre GCSEs but I set myself high goals for A levels. Money doesn't have to put you off – my mum works in an office and my dad's a gardener.

> Jamila: I set myself a target – Royal – I come from an under-privileged area.
>
> MAS

The power of the individual to overcome this disadvantaged positioning was referred to in these accounts. Through her own aspiration and 'setting herself a target', Jamila, one trainee ambassador, explained how she managed to defy expectations of pupils from underprivileged areas.

This discourse of individualism obscures the reality facing young people, namely that their futures are very much constrained by existing social structures. It also has the negative consequence of creating a corresponding discourse of individualized blame, with young people being held and seeing themselves as responsible for their own failure to succeed within the HE system (Ball *et al.*, 2000; Evans, 2007).

Medical days: missionaries working for the under-privileged?

The outreach work of the student ambassadors on the MAS was partly funded by a 700-year-old livery company based in the City. The charitable origins of the scheme and how it was funded was reflected in how the students on the scheme were discursively constructed. The outreach manager's account reveals that the ambition of the scheme is to enable a few young people with ability, from generally low-achieving schools, to progress onto a degree course in medicine:

> If you're at one of the crappy schools, staff can be wonderful and you can get three Bs compared to someone at St Paul's who gets four As. If you do that you're a genius. It's about levelling the playing field – it's not widening participation in the sense of targeting certain groups – it's just, you can be a doctor despite exam results.
>
> MAS outreach manager, Royal

This again reflects dominant discourses of individualism; that there are a few special individuals, who, because of their natural ability, hard work and determination, and with a little help, can succeed in accessing areas that are traditionally the preserve of the middle classes. This discourse, like the livery company itself, can be traced back to Victorian times.

A related discourse repeatedly drawn on by ambassadors was that of underprivilege. This was used both with reference to themselves and in relation to the pupils that they were working with. Indeed, one ambassador explained her own experience at school had motivated her to want to work with pupils as an ambassador:

> Remi: 'Cause we didn't get, like, in my school, no-one came in to talk to us about uni life and in my family I am the first person to go to uni so my mum couldn't advise me, my friends couldn't advise me 'cause they're in the same boat, so doing this for the kids as well, you're helping them, you're giving them more of an insight than what their teachers or what their parents can give them.
>
> Medical day

Another ambassador suggested this shared background enabled the ambassadors to challenge what pupils perceive to be possible for them. She describes how pupils were shocked when they found out that she went to the same school as them or lived in the same area:

> Candice: Sometimes it's good when you have someone who went to the same school as you or are from the same borough because they're like, obviously you got here because, you know, you're … and you're like, well, 'I went to the same school', and they're shocked, they're like, really, 'no you didn't'. So you just show them that you can do it as well and maybe, like, obviously, like you were saying you were the first one in your family to go, they see, okay it is possible.
>
> Medical day

One ambassador described how she, like the school pupils she was working with, came from a 'lower-privileged background', how this shared identity 'inspires them'. She went on to explain how it is 'expectations in your head' that constrain pupils from progressing:

> Chanelle: Yes, you don't have to come from an upper-class background or a grammar school to get to university. You can come from where they are coming from. There's no real boundaries apart from your actual expectations in your head, I think. It's like, if you think you won't be able to make it, then that's going to limit you in where you're going. If you think, 'I can do this, I can achieve what I want to achieve', then that will give you inspiration to go, and if there is someone telling you, you know I came from where you come from; I came from a lower-privileged background and I'm here, it inspires them.
>
> Medical day

One ambassador's account of his work with pupils drew on religious terminology, suggesting that ambassadors were there to 'give them hope' and to encourage pupils 'not to give up', and that they are 'preaching that message'. He even refers to aspiring to study at Royal as aspiring 'to a higher state':

> Dan: The people we tend to speak to because they come from certain boroughs, like they do find it hard to believe that they really can achieve so much so I think our role is, like, encouraging them, giving them hope that you know, don't give up and preaching that message, don't give up. You know you can do what we want to do; aspire to a higher state and aspire to have ambition – I think that is the key role that we do.
>
> Medical day

One ambassador drew on discourses relating to real poverty to describe the plight of school pupils with 'not enough food on the table':

> Chanelle: And as you see on the news now that graduates aren't getting anywhere or there's not enough funding for university places, you know, all that stuff and the whole economic climate – even those you may think are not aware of it, they may be aware of it, they may think, well, my mum's lost her job or, you know, there's not enough food on the table, but if I can get somewhere and make myself a better person you know, by getting to university, why not?
>
> Medical day

Dan responded to this by again drawing on religious terminology when talking about the need to spread the message to pupils that they can achieve their ambitions and that it's not 'too hard':

> Dan: And yet that is so too hard – it's too hard – that's what kids, like – that's what you are afraid of – 'it's too hard and I can't do it', and they'll go, 'I can't do something like Simi was saying that I'll just do something that I probably can do'. Well like, if you can, encourage them and get that message across.
>
> Medical day

In order to understand the passionate nature of these accounts of their work, it is necessary to examine the action orientation of these ambassadors. An ensuing discussion about ambassadors' own experiences at the schools in the boroughs where these students come from provides some insight into why they are so motivated. The ambassadors discussed how little support they received from teachers and how little information, advice and guidance they received when they were at school. They also discussed the impact this deficit had on them and their perception of how this affects the school pupils that they work with:

> Remi: 'Cause in my school we didn't get careers advice until it was nearly too late. If you haven't done the GCSEs then there's nothing you can do. If you haven't done the A-levels then you're stuck. So my careers advisor told me to work in social services instead when I told her I wanted to do medicine. She said it will be easier if you go into social services. I was like, 'well you are supposed to be advising me how I can progress, not telling me to do something completely different', and at the age that I was, if I didn't find out myself what A-levels I needed, what GCSEs I needed, I would have been completely lost. So I think you need careers advice from Year 8 or Year 9 so you know what GCSEs you want to pick.

> Falak: I remember teachers, when I was in secondary, they never thought that I could reach anywhere so far because I was quite … I used to pick up things less easily than others. I was quite slow at school and teachers used to always tell me, you shouldn't go into further education, go and see if you can do something else. Even at sixth form I was told there's no way you can actually get into medicine; see if you can find backup choices as well and I was pressured into choosing other things rather than doing medicine. I think if it wasn't for this programme I don't think I'd be doing medicine now. If I wasn't … actually had my friends or support with me I don't think I would actually have got into medicine, so I am glad that I actually had that support.

> Simi: At my college, my biology teacher said, 'you're not getting into medicine', you know, regardless of my grades that I had; regardless of my determination, just told me, 'you're not getting in'. I was like, 'okay, you're supposed to be my teacher; you're supposed to be supporting me, you're supposed to be the one I look up to, to help me get my grades but yet you're telling me I'm not getting it', so when I got in I was like – ha [laughter]. So it's like you need support from those around, not only your family 'cause your family can do so much in your academic life, you know the teachers are there to get you or get you thinking where you want to go.

> Adi: Well, I found that sometimes when teachers put you down and expose you to, like, you want to prove them wrong.

Remi: But then if you didn't have strong self-esteem, 'cause you'd always been told that you'll never be able to do it, then it's hard to be able to fight against that, so I think it's important for children in schools to have mentors or these schemes where they can speak to older people and know that there is a possibility: 'I can do this if I just put the work in, if I just set my mind to it and a lot of the time.' They don't get that in schools and that's disappointing.

Falak: Yeah, because when I was really young I was quite frightened just thinking about university. Even at college I was really frightened at a young age to think about anything – further education. I thought I would probably end up getting a job or something ... My secondary school didn't do much because the careers, again like Remi mentioned at her school, they don't have much.

Medical day

These accounts reveal a genuine issue regarding the availability of careers information advice and guidance in these inner-London schools. It is a concern that the provision of careers information, advice and guidance appears to have deteriorated further, with the House of Commons Education Committee (2013) pointing to a potential increase in problems following the withdrawal of the Connexions service and move to give schools this responsibility without additional funding. As one ambassador pointed out, decisions made early in young people's learning careers have serious repercussions: 'If you haven't done the GCSEs then there's nothing you can do.' Another ambassador commented on how important it is for 'children in schools to have mentors or schemes where they can speak to older people and know that there is a possibility'.

Common to the stories of all these ambassadors is the complaint that their teachers were not supportive of their ambition to study medicine. However, the student ambassadors did not attain high enough grades to acquire places on standard medical degree programmes, which may account for teachers' circumspection. Ambassadors' expectations that their teachers should be able to offer careers advice ignores the fact that teachers are generally not well informed about different routes into careers. Despite this, the ambassadors' suggestions that teachers 'put you down and expose you' are a depressing comment on the culture in some schools. These accounts also suggest these students' reliance on their peer group, 'friends to support me' and the lack of relevant advice from family, 'cause your family can [only] do so much in your academic life'.

Overall, these accounts from Royal overwhelmingly reveal a sense of deficit, an awareness of lack of privilege that I did not encounter at Bankside, though ambassadors at both universities were from similar geographical areas and backgrounds:

> Remi: 'Cause we didn't get, like in my school, no-one came in to talk to us about uni life and in my family I am the first person to go to uni so my Mum couldn't advise me, my friends couldn't advise me 'cause they're in the same boat.
>
> Medical day

One ambassador's explanation that he was 'frightened ... thinking about university' reflects this, as did other ambassadors' accounts of lack of support from their schools and teachers. One specifically explained that 'you don't have to come from an upper-class background or a grammar school' to go to university and how she came from a 'lower-privileged background' herself. This provides some insight into the awareness that these ambassadors have developed of their own positioning in relation to their wealthier and more middle-class peers at Royal. As Reay et al. (2005: 33) suggest, the education system 'valorizes middle-class ... cultural capital'. For ethnic-minority or working-class students from inner-city schools with no history of HE, it is clear that existing at a medical school like Royal, where the overwhelming majority of pupils are from established middle-class backgrounds, is done at some psychological cost (Reay et al., 2005). These students' learner identities (Reay et al., 2009), though important, also appear relatively fragile, as they were not identified as high fliers in their schools and achieved lower A-level grades than their peers on the standard degree programme. They appear to live with the fear of being 'found out' to be 'inferior, less cultured, less clever than the middle classes' (Reay, 2001: 343). Their positioning at Royal resulted in ambassadors identifying closely with the school pupils they were working with and contributed to their almost evangelical zeal in encouraging pupils from similar backgrounds to progress to university, to study medicine and to study medicine at Royal on the MAS.

Marketing subject identities

Marketing discourses were evident at every level of ambassador work, from government policy through to the practices of organizers within individual HEIs and of the student ambassadors themselves. The ideology of the marketplace operates as a regime of truth (Foucault, 1980) dominant in the contemporary UK HE system.

How the student ambassadors were used within each institution reflects the stratified nature of the HE system. The ambassadors were used as recruitment tools to attract appropriate students to these universities and were primarily marketing their own institutions. The outcome of their work is likely to be the maintenance of the status differentials of their institutions and the perpetuation of existing patterns of recruitment to those universities. However, ambassadors' positioning as 'hot sources' of information that can be trusted by younger pupils may be threatened by student ambassadors being too flagrantly positioned as marketers of their universities and pupils as consumers.

Positioning student ambassadors as marketers of particular subjects and courses may, in some contexts, be effective. While it seems likely that levels of interest in subject areas are formed early, and this relates closely to the extent that pupils identify with subjects and to their perceptions of who or what particular subjects represent, student ambassadors appeared effective in reinforcing and reaffirming subject interests and orientation to their study at university. However, in positioning ambassadors as marketers of institutions and particular subjects, school pupils are correspondingly positioned as 'individualized, self-directed consumers of learning' (Malcolm, 2000: 20). While it may be influential, it is unclear whether in this marketplace the information that student ambassadors provide is accurate or valuable in terms of facilitating informed choices of subject to study.

The dominance of marketing and related discourses of professionalism at Bankside and charity and deficit at Royal also reflect the dominance of discourses of individualism. HE is represented by such discourses as being for the benefit of the individual. The implication is that it is up to individual pupils to raise their aspirations and aspire to university generally and to elite universities such as Royal in particular. These individualized discourses reflect neo-liberal discourses that have gradually overwhelmed and subsumed post-war discourses relating to the social value of HE. These neo-liberal discourses of individualism, marketing and credentialism have impacted widely on the learning outcomes of WP activities.

Chapter 6

Learning practices and identities

An analysis of ambassador work reveals patterns in pupil learning which reflect the type of activity engaged in, the pedagogical approaches used and how these influence the relationships between ambassadors and pupils. During events and activities, ambassadors are differently positioned as teachers, careers advisors or even as friends. This positioning affects the nature and content of the learning that occurs and the balance of informal and formal attributes in these learning contexts changes the nature of the learning taking place (Colley, 2005; Gartland, 2014).

Stakeholder interests, including those of subject experts, dictated the focus and nature of the WP activity. Engineering and medical specialists contributed to the design of activities for the AEP and for the MAS, which was generally not the case in other WP work that I observed (WP coordinators at both institutions described difficulties in engaging academic staff). In the UK and internationally, there have been significant moves to problem- or project-based and experiential learning in both medicine and engineering (Albanese and Mitchell, 1993; Sainsbury, 2007; Arlett, *et al.*, 2010; Northwood *et al.*, 2003). These pedagogical approaches are generally viewed as more effective in engaging women and more diverse students (Arlett *et al.*, 2010; Boursicot and Roberts, 2009) and were reflected in outreach activities organized as part of the MAS and AEP, though there are significant differences between HE teaching and outreach work with its more general aspiration-raising focus.

Attributes of formal learning and their effects

In all contexts the ambassadors were employed to work with school pupils but in different capacities, depending on the views and objectives of stakeholders and organizers. None of the activities were set up by the ambassadors themselves. As Colley (2005) identifies, an understanding of this wider context is vital when considering any learning that is taking place.

The differences between 'learning situations' have been theorized in terms of formal and informal learning. Colley *et al.* (2003) suggest that, in practice, 'elements of both formality and informality' can be found in every learning situation and that, instead of these being described as formal or informal, formality and informality should be identified as 'attributes' of these situations. They use the term 'attributes' advisedly, both to suggest that learning has many attributes, and to highlight that the labels are 'attributed' by writers and that learning is not either inherently formal or informal. Colley *et al.* (2003: 30–1) outline four main groups of informal learning attributes: 'process', 'location and setting', 'purposes' and 'content'. These provide a useful framework for exploring different learning contexts.

Teaching the syllabus

I have described the emphasis on providing customer satisfaction during a Bankside meeting with organizers and stakeholders in student ambassador work. One clearly identified group of consumers was teachers in schools. A borough coordinator for Aimhigher commented on the 'growing requests from schools for mentoring and teaching' and suggested that student ambassadors should contribute to both aspiration raising and 'raising the C/D borderline'. This focus on raising attainment in schools, and particularly on raising the levels of achievement of groups of borderline C/D pupils, especially in maths, has become a pressing focus in schools (Williams *et al.*, 2010; Gillborn and Youdell, 2000). The pressure of school league tables has translated into the types of request made by schools for ambassadors, and discourses of credentialism have then been taken up and practised by organizers of WP activities, functioning as another regime of truth in the context of outreach work.

At one south east London school student ambassadors from several universities, including Bankside and Royal, were involved in working with Year 11 pupils during maths workshops. These sessions ran after school two days a week. The learning taking place during the workshops had notably more formal attributes than other contexts considered in this study. The ambassadors were working with pupils on exam papers, where in terms of 'content' their learning is propositional (either true or false) and outcomes are rigidly specified. In terms of 'purposes', the learning is the 'prime and deliberate focus' of the activity and is 'designed to meet the externally determined needs of the exam board' (Colley *et al.*, 2003: 30–1), and in terms of 'process', the approach taken by ambassadors in this context is didactic and the assessment of the learning is the formal GCSE exam. The location of the sessions in a school classroom is another formal attribute in this context.

A teacher was present during the sessions and student ambassadors circulated among pupils, helping them with questions from GCSE exam papers. One of the teachers who supervised the workshops explained what she thought the ambassadors were there for:

> Well, my understanding of it is that they're coming to a maths club, an organized maths club, in order to support students in their goal to get at least a grade C in their GCSE, and that by working with … our students working with people that are closer to their age group or closer to their kind of point in life … it might be easier for them to access the work, to kind of walk into a mentoring scheme rather than a teaching type of environment. So it's like a change of environment from what our students are used to, but also that they should be maths specific, that they probably are maths students that are coming in to share their skills.
>
> Teacher, maths workshops

The formal purpose – to support pupils in their goal to get 'at least a C' – is very clear in the description of the ambassadors' work. The teacher's understanding of the benefits of having university students in to work with the pupils is less clear, though she identifies their proximity to pupils in terms of age and 'life point' and their maths expertise as important, and outlines a 'change in environment'. This type of student tutoring has become popular in the UK and is identified by Colley (2005: 32) as an attempt to 'increase informal attributes of learning in situations traditionally regarded as formal'.

Pupils admired ambassadors' knowledge of maths and appreciated the commitment of some of the ambassadors to helping them. However, their criticisms are revealing. Pupils variously described ambassadors as 'too fast', as having 'done the work for me' and as 'going through the whole topic' and 'wasting' time. Pupils were particularly critical of one male ambassador from Bankside who they describe as 'not that good' at explaining. His explanations were criticized for being unnecessarily 'complicated' and he was also perceived as lacking in commitment:

> Yvonne: His explanation is not that good.
>
> Clare: You said he was quiet.
>
> Yvette: Yeah, he speaks quite low and I can't hear him.
>
> Clare: Go on – explain what you mean by his explanation is not that good.
>
> Yvonne: Like, he'll always do the longer way – 'cause there's always an easy way to do a certain thing but he always does the

Learning practices and identities

long way and he teaches us things that we don't really need and we're not going to remember in exams.

Bim: Yeah, he showed me this complicated stuff.

Yvette: The way he comes in, like they're forcing him to come in [laughter], he just comes in like they are forcing him, like, when he comes in he just sits down and waits for a student – he doesn't get up and say, do you need help, he just sits there waiting for people to come ...

Bim: In the other room he sits at that table near the back ...

Yvonne: Like he's lazy – I don't think you guys pay him to do what he's doing – it's not enough for him, sorry [laughter].

Yvette: To motivate him.

Janine: At least he comes in sometimes, most of the time. He's always here so that's a good thing.

Yvonne: You need to reduce his hours [laughter].

Maths workshops

During the maths workshops, ambassadors were positioned in formal didactic roles similar to teachers. Pupils' discursive constructions of the ambassadors' work reflect this. They repeatedly comment on the quality of the explanations given by ambassadors. They also draw on discourses relating to the professionalism of ambassadors and expect ambassadors to take the lead in interactions. Ambassadors were generally viewed by pupils as inadequate substitutes for real teachers, though having *any* help was seen as better than having none:

Clare: If they weren't here what difference would it make?

Yvonne: We would have to wait for the teachers and it would be really hard to get a teacher 'cause everyone would need them.

Janine: Because they are always busy.

Bim: And obviously they can't help you with everything 'cause they have to do their business as well.

Clare: So they ...

Leticia: But the teachers are really useful as well. I think they are better than the mentors.

> Yvette: Yeah, 'cause even though I have worked with a mentor, they may be good at maths – at the end of the day I still go back to the teacher and ask her this and ask her that and she went – Adam I should get a lot of … .
>
> Bim: Because you know, they do a different type of maths at university so they might show you university style and then you just want to know GCSE style.
>
> Yvonne: I used to argue with Adam – I would say, that's the way to do it – he'd say, that's another way.
>
> Yvette: Adam [laughter].
>
> Bim: You were arguing with him?
>
> Yvonne: I knew a formula that, what's it called – where you do that table thing?
>
> Bim: The green method.
>
> Yvonne: Yeah, the green method, I would say, that's the way you do it and he was like, 'no, you have to do that', and I went and told Miss M and she said it was the green method.
>
> Maths workshops

Teachers involved also commented on the problems posed by positioning student ambassadors as teachers of GCSE maths when they have had no experience or training:

> Rachel: I would never ever start a lesson without a clear objective, but we started the scheme with no clear objective for ourselves, our students and for the students that were coming in, which is possibly why I don't think it's been as successful as hopefully it will be next year.
>
> Clare: So go on, what do you think the issues have been this year?
>
> Lucy: One of the main issues which may well be – well, no, I think it is closely related to communication, is the fact that we're not convinced that all of the undergraduates have the skills required – maths skills I am talking about now, not just communication and imparting knowledge skills. I don't think they have the maths skills required, mainly because, although they are closer to that point in time in their life of having taken their GCSE, they've moved on

Learning practices and identities

to do much more complicated things and the GCSE syllabus is such that very little of it is carried on in a pure maths degree or an applied maths degree, so I think for me something needs to happen so the mentors need to have the materials well in advance. I think preparation on our side wasn't as good as it could have been.

Teachers, maths workshops

The teachers' accounts relate closely to those of the pupils. Positioning student ambassadors as teachers simply because of their subject expertise in maths is evidently problematic. The ambassadors were taking up this positioning to various degrees of success and enacted being GCSE maths teachers. Far from 'increasing informal attributes' in this learning context, ambassadors embedded existing formal attributes. However, this practice, as Rachel suggests, can impact negatively on pupils' subjectivity and on their confidence and sense of self-efficacy in maths, with pupils at times unable to understand the ambassadors' explanations.

> Rachel: … they [ambassadors] start going through a maths paper and they do all the easy stuff and they are thinking, oh this is great and then they continue for the next hour to try and teach this grade C student A* grade stuff and that's just never going to happen … and actually the student's confidence might get knocked by not really understanding these things they will never have learned in their lessons because we're never going to teach it to them, but that's not the ambassador's fault, that's our fault because they need to be trained on that kind of stuff.

Teacher, maths workshops

Lucy also suggests that the experience was not always positive for the ambassadors themselves; this may account for the fact that a number of ambassadors only worked with the group on one or two occasions (I attended the club for several weeks and did not observe any students from Royal):

> Lucy: I think that's one of the hard things being an ambassador, coming in as a peer mentor … you may have no other experience of school other than when you were at school, so you are not used to being proactive … and you've got a lot of teenagers sat in this room who are talking to each other and I imagine it would be quite intimidating.

Teacher, maths workshops

The ambition to maximize numbers of C grades among pupils in maths may well be laudable, but the effectiveness of ambassadors in supporting this aim without relevant training appears to be questionable. It is also important to question the learning about maths that this approach reinforces. In Williams *et al.*'s (2010) view, a narrow focus on examination practice presents maths as having 'exchange value', with pupils only focusing on their learning for exams in order to gain access to the next stage in their schooling. They suggest that this focus on the exchange value of maths promotes identities among pupils as surface learners rather than as 'users of mathematics' who are confident and engaged mathematicians. These data reveal that embedding STEM ambassadors in existing contexts with many formal attributes is unlikely to support engagement with maths as a subject.

The engineering camp presented a very different learning context. Pupils were staying at a campus university (though no students were in residence at the time) and were involved in practically oriented engineering activities including building eco houses and model aeroplanes. This reflects the more experiential learning approaches increasingly found in HE. However, the ambassadors' role as supervisors responsible for ensuring that pupils completed projects within the available time again positioned ambassadors in particular ways. As with the maths workshops, the help of ambassadors was generally welcomed by pupils, who identified less time waiting for help from teachers as a benefit. However, this help was sometimes viewed as prescriptive. One pupil complained that 'there's always someone telling you what you should and shouldn't do'. This discursive construction of ambassadors as being there 'telling' pupils what to do relates to the propositional content and didactic process which are seen by Colley *et al.* (2003) as attributes of formal rather than informal learning:

> Janine: I worked with one of them making a house – it's been ok. It's the same as school but there's more help – there's someone to ask, but there's always someone telling what you should and shouldn't do – sometimes you want independence.
>
> Cantrice: ... if they weren't here – half and half – they wouldn't be there to assist you and help you with ideas, but it would be better 'cause there wouldn't be so many people to tell you what to do.
>
> Mark: ... it would be boring without them – there would just be lots of people saying, 'what do I do now'?

Nick: Compared to school there are more of them (supervisors) concentrating on one thing so you get there faster and better – you can't stay focused on one thing at school.

Engineering camp

Pupils were given unstructured long stretches of time to undertake activities. While some were motivated and engaged in completing tasks, others were left with little to do and appeared bored, expressing frustration that they were 'doing the same thing every day'. Conversely, then, the lack of structure within the activity, which is often viewed as an attribute of formal learning, contributed to pupils' lack of engagement.

During the meeting about ambassador work at Bankside, the outreach manager suggested that ambassadors should take on a 'leadership role in developing work' and should be able to present to 'a group of kids':

I'd like student ambassadors to take more of a leadership role in developing work we do in schools and colleges … . You get ambassadors who in front of teenagers get carried away with being too jovial … . I think ambassadors should be able to stand up in front of a group of kids … I think they should have the potential to be able to stand up and talk to a group.

Outreach manager, Bankside

Positioning ambassadors as responsible for the learning of pupils reflects the trend discussed by Colley (2005) for students acting as learning mentors in schools. However, this positioning can serve to undermine social relationships between pupils and ambassadors.

During the G&T summer school at Royal, ambassadors spent several hours on the penultimate day working with groups of pupils and helping them prepare a presentation. They discussed the challenges and contradictions posed by this responsibility for pupils' learning:

Munira: It's interesting, isn't it, because the dynamic of my group is really quiet, like, they hardly said anything … . I've been more like, 'come on, come on guys, right, you do some research on this, you do some research on that' … . Yeah, it's difficult …

Clare: So, I mean, in terms of friend/teacher [*both terms used by ambassadors earlier in the discussion*] – where does that put you?

Munira: I guess it really depends on the situation, doesn't it, 'cause if there's something that has to be done, the onus is kind of on the ambassador to kind of make sure … we have to make sure that they are able to do it ... so I guess it's kind of motivating and being the motivating force in whatever capacity you feel is most appropriate.

Carla: Sarah [WP officer] said at the beginning, 'they are young adults – we should not call them kids and stuff because they would be quite offended – like 15/16 year olds being called kids 'cause they think themselves adults so we do treat them' ...

Than: I feel like I don't treat them that much differently.

Munira: I think it is different like, we treat them as if they're our age, but at times they're not therefore you need to sort of be like right, well – come on, do something.

Carla: They do need to be pushed.

G&T summer school

The ambassadors' discursive constructions of their interactions with pupils reveal that they are positioned as teachers, allocating tasks and organizing their groups – 'come on ... right you do some research on that'. Munira describes how 'the onus is on the ambassador' to ensure that the pupils are 'able to do' the presentation. There are attributes of formality in the prescribed 'externally determined' purpose of this work. Munira feels under pressure to ensure that the pupils are prepared and identifies a central problem when she comments 'we treat them as if they are our age but they're not'.

Pupils' responses to the positioning of ambassadors as responsible for their learning are also worth noting. A group of three boys on the engineering camp interpreted the ambassadors in the light of familiar adult roles in schools. Ambassadors were consequently seen as there to support the less able, with the boys identifying themselves as not needing such support. This view of ambassadors reflects the pupils' experience of classroom practice where teaching assistants are often used to support particular groups of pupils who have a special educational need (SEN). It may be that these responses from male pupils also hint at different gendered responses to the work of ambassadors in this context (Gartland *et al.* 2010):

Nigel: I haven't worked with any of them.

Bill: they only help you if you need it.

Engineering camp

The data illustrate how activities where ambassadors are responsible for pupils' learning, particularly for curriculum learning, are problematic, creating social distance and conferring formal 'didactic' attributes on the ambassador–pupil relationship.

Managing difficult behaviour

There were tensions around the authority of the ambassadors and the pupils' behaviour. Supervising the learning of relatively large groups of pupils during the G&T summer school presented tensions for ambassadors. On the engineering camp, where ambassadors were clearly positioned as supervisors directing and managing pupils, this noticeably affected their ability to develop social or even effective working relationships with pupils.

The ambiguity in ambassadors' positioning as authority figures was evident during the meeting about ambassador work at Bankside. The ambassadors' responsibility for managing the behaviour of groups of pupils was reflected in the suggestions made by the lead ambassador present at the Bankside meeting. He identified the need for training in classroom management:

> Would you do something like classroom management ... everyone at some point is going to have to deal with a number of kids ... we had an issue recently with how they handle kids – maybe they need some training on how to handle that ... to do events, we want people to have classroom management skills.
>
> Lead student ambassador, Bankside

The borough coordinator outlined a tension being that, when ambassadors go into schools, they may be left in charge of groups, although this is not supposed to happen:

> Someone in education could come in and do a session on management, but it's not supposed to be them in charge of a class ... they perhaps need to know when to step back and get members of staff who think it's free time and just sit back and do nothing.
>
> Borough coordinator

This tension was also evident in the advice given to ambassadors during training at Royal:

> You are an ambassador, not a teacher or lecturer. Let school staff do their jobs and manage the situation where appropriate ... you do not have to deal with disruptive pupils – some will be anti-going to university.
>
> WP officer, Royal

At Bankside, part of the ambassador training was a session on 'conflict avoidance' run by someone who trains police officers. This focus in the training aimed to provide ambassadors with help in dealing with being in an authoritative position. The outreach manager explained he felt this was useful as ambassadors often struggle in these contexts:

> When student ambassadors are in groups of 14 year olds, they tend to move into teacher mode and try to be disciplinarian or start to be sarcastic.
>
> Outreach manager, Bankside

The ambiguity over authority was also reflected in ambassadors' own accounts. One Bankside ambassador commented during the careers session that 'I try to make it clear – I am not the teacher – but still they have to listen'. These tensions were evident in the ambassadors' work at the G&T summer school at Royal. Munira's account outlines a tension between the need to 'speak on a level' in order to 'fit in well with' pupils and 'make bonds', and the different working contexts where 'you're in a situation where you have to tell them to quieten down. She refers to the need for 'authority' and having 'a teacher role' and the conflicting positionings of friend and authority figure:

> Munira: I kind of speak on a level with them 'cause then they feel more comfortable talking to you and then you can fit in well with them and it's easier to make bonds, but obviously when you're in a situation where you have to kind of tell them to quieten down, you take authority, then you have your authoritative voice like, 'guys, come on quieten down', this sort of thing – then you become like a teacher role but usually we just go and speak to them like …
>
> Than: As a friend.
>
> Marvin: I think we're more friends than teachers, but there's times where we need to be – have the same authority as a teacher, I think.
>
> Carla: I do think they still respect us even though they feel they can talk to us … they definitely …
>
> G&T summer school

While the cohort of pupils on the G&T summer school was described as very well behaved, this was not always the case. The ambassadors described the pupils on the Aimhigher WP Easter School, who are not G&T and are selected on quite strict targeting criteria in terms of their backgrounds, as being 'a bit rough'. This discursive construction of 'rough' pupils resonates with those

Learning practices and identities

found in Taylor's study, where ambassadors' experiences of attempting to manage difficult pupil behaviour in classroom contexts contributed to an entrenched sense of difference between ambassadors and pupils:

> Interactions between university and school pupils at times pointed to the divergent attitudes, hopes and behaviours, as opposed to seamless peer 'sameness'.
>
> Taylor, 2008: 159

The positioning of ambassadors as authority figures responsible for pupil behaviour was taken up by both pupils and ambassadors on the engineering camp. During conversations with pupils over the two days it emerged that many did not differentiate between the help provided by ambassadors and that provided by teachers and other adults present. Pupils repeatedly referred to all adults as supervisors.

Pupils on the engineering camp identified the work of the ambassadors as being predominantly disciplinary, with a group of three specifically complaining to me about their 'supervisor'. Despite the fact that they were well matched in ethnic and gender terms, and the ambassador was viewed positively by pupils in other learning contexts, here she was viewed with hostility:

> Janine: Some of them need to change their attitudes towards us – they can be so rude. My lead supervisor, she's so moany. In the morning she complains and moans 'cause I take a long time to do my hair.
>
> Carly: I don't appreciate that – knocking on your door – 'time to get up!'
>
> Janine: She barged into my room.
>
> Engineering camp

One ambassador explained that the engineering camp required more focus on discipline than other activities she had been involved with as an ambassador:

> Gill: We go into schools and work on events and it's the same but this is more so, as you're with them 24/7 … . In school there's someone else to stop them leaving the classroom.

She appeared reluctant to take on the disciplinary work of teachers:

> Gill: At the same time I couldn't view myself as a teacher – I would hate to do that – when they call me 'Miss' I tell them to call me by my name – drop that barrier.
>
> Engineering camp

Several ambassadors involved in running activities during the engineering camp expressed a desire to distance themselves from the disciplinarian work of teachers, but felt they were positioned as teachers by the pupils themselves. One organizer, in constructing herself as 'other' than teacher, discursively constructed disciplinary work as the responsibility of the ambassadors:

> They think we're teachers. I'm not a teacher – I'm not there to tell them what to do – that's the ambassadors' job.
>
> Engineering Trust employee, engineering camp

One of the ambassadors talked explicitly about how she felt that the work she was doing was 'like a teacher' or another formal role found within schools, that of teaching assistant:

> Gill: It's like being a teacher … a classroom assistant.
>
> Engineering camp

These discursive constructions of ambassadors' work were echoed in the account of a teacher who was accompanying a group of girls on the course. She suggests that the ambassadors need assertiveness training so that they can 'take on a teaching assistant role'. This teacher's own positioning in school informs her view of the work ambassadors should be doing:

> They may need assertiveness training – the kids think they're at the same level as them – they need training so they can take on a teaching assistant role.
>
> D&T teacher, engineering camp

Stakeholder interests, particularly those of teachers and schools, impact on the positioning of ambassadors. Outreach activities need access to school pupils (Gartland, 2009; HEFCE, 2010a). Organizers of activities within universities inevitably attempt to accommodate the wishes of teachers in order to reach pupils. In the case of Aimhigher partnership, working with schools and other stakeholders is written into funding agreements. In these various contexts, these interests position ambassadors as supporting the work of teachers and schools; they are even understood by pupils and teachers in the light of familiar adults in schools, but this can be an uncomfortable positioning for ambassadors.

It is also important to consider what subject-specific learning ambassadors facilitate. During the maths workshops, the emphasis on 'mathematics for exchange' supports and reinforces the credentialist regime of truth operating across education and particularly in secondary schools. The workshops and the engineering camp are both contexts where the learning process and content of activities, as well as the location of the maths workshops in a school classroom, dictate that ambassadors take up formal didactic positions in relation to pupils.

Attributes of experiential and informal learning and shared learning identities

The activities studied could be placed on a continuum of their formal and informal attributes, with the maths workshops at one end and Train Tracks at the other. The summer school at Bankside, the medical day and medical afternoon at Royal also had comparatively more attributes of informal learning. If we consider process, all these activities were more negotiated and there was no planned formal assessment or pressure on ambassadors to ensure particular learning outcomes. There are no predetermined learning objectives, curriculum or external certification (Colley et al., 2003).

Working collaboratively (Beckett and Hager, 2002) and supporting pupils with 'uncovering knowledge derived from experience' (Colley et al., 2003: 31) rather than being prescriptive was important to the development of positive relationships between ambassadors and pupils in these contexts. Student ambassadors listening to pupils, allowing pupils to lead and negotiate their own learning and to explain their views without interruption, were important informal attributes of these learning contexts and facilitated warm and open relationships between pupils and ambassadors.

A key difference in how pupils were learning during Train Tracks and the maths workshops was that the former provided pupils with practical experiential learning. According to constructivist thought (Piaget, 1966; Vygotsky, 1978; Schon, 1983; Kolb, 1984; Boud et al., 1993; Boud and Miller 1996), experiential learning is 'concrete experience, reflective observation of experience, abstract conceptualism and active experimentation' (Colley et al., 2003: 56) and is 'supportive and enabling rather than intellectually prescriptive' (Kyriacou, 2009: 53). This experiential approach was advocated by Lord Sainsbury (2007), who outlined the need to develop 'experience led' engineering degrees. This approach to engineering education is gaining hold in UK universities (Arlett et al., 2010) and is now well established in medical education where problem-based learning has become ubiquitous (Albanese and Mitchell, 1993). These pedagogical HE strategies have filtered down into some of the outreach work considered here and this is likely to have been supported by the input of subject specialists involved with the AEP and MAS.

Pupils worked with ambassadors on subject-related experiential activities across the different learning contexts. During the summer school at Bankside, the pupils worked in groups with ambassadors programming and building robots. During the medical afternoon and medical day pupils worked alongside ambassadors using medical equipment and experimenting with their DNA. During Train Tracks, pupils and ambassadors worked practically together building train platforms and tracks. Pupils and ambassadors were provided with a set of real-world problems (connected to tracks and platforms) and pupils had to develop a cost-effective solution which involved drawing on their science, design and maths knowledge. This practical experiential learning was also present during the summer school at Royal, with pupils using medical equipment and completing other practical work, as well as during the engineering camp. Experiential learning appears to be an important strand at most events and activities, but other attributes of formal and informal learning varied, including the balance between experiential learning and more didactic practices.

Practical, experiential and collaborative working

During the summer school at Bankside, pupils were keen to identify the 'authority' of the ambassadors as a key difference between them and their teachers. One group of pupils explained that the ambassadors were 'not … allowed to tell us what to do':

> Michael: … and they take part in the activities as well, but they're not really allowed to tell us what to do so …
>
> Sarah: They hint, yeah.
>
> Michael: They hint – that's how they keep us on task.
>
> Clare: By hinting.
>
> Bankside summer school

Sarah suggests that ambassadors hint rather than tell the pupils what to do. The pupils discuss how this lack of telling enables them to learn more independently, using their own initiative and allowing for individual interpretation, which 'allows everyone to do things differently':

> Michael: I think it is good because then the course wouldn't be as challenging.
>
> Joe: And we have to think for ourselves.

Learning practices and identities

Sarah: Otherwise we would just be using what they say.

Joe: What they know – so we use our initiative.

Michael: I think it's like ... the learning environment is like ... we have fun whilst you're learning and allows everyone to do things differently – you see how people can interpret something that they're told to do differently to everyone else.

Bankside summer school

Pupils discussed how lack of understanding is treated differently by ambassadors than by teachers in school and how the lack of pressure and an emphasis on fun make the learning easier and more accessible:

Joe: We don't realize that we're doing something that we're learning at the same time. It's not like in school where, it's, like, if you don't get something, you feel like it has to be drilled into your head. It's, like, you understand it easier.

Michael: It's that there's humour in learning the stuff – like half of us here – maybe some of us have chosen engineering, but some of us got put into engineering because we didn't get our first choice so it's like – we're doing stupid things and we don't know what we're doing but it's made ... funny.

Bankside summer school

These accounts resonate with Colley *et al.*'s (2003) description of informal attributes of processes, content and purposes in learning contexts and with the 'active experimentation' through 'experiential learning' that they describe. The pupils discussed how the relationship with ambassadors was 'easy' and 'comfortable'. This ease appeared to be closely connected to the learning context and the positioning of ambassadors within it.

Resonating with pupils' accounts, the ambassadors also explained that they were 'not like a teacher':

Clare: If you were to compare yourself to another adult in the students' lives, who do you think you would be?

Qadira: ... just in between – like the assistants, or just there to help – like, not the teacher – so they can kind of get away with stuff with us around that they wouldn't if a teacher was around – so I don't know – just a role model I suppose – just a young girl – they still call us 'Miss'.

Clare: So somebody in between?

Adam: In between, a teaching assistant.

Qadira: They know the boundaries – they do know the boundaries, but they seem to be more comfortable with us so still keeping the …
Bankside summer school

Like the pupils, Qadira describes the difference as being rooted in the ambassadors' lack of authority, 'they can get away with stuff with us'. Like the pupils, Qadira suggested a difference between the ambassadors and teachers as being that the pupils 'are more comfortable with us'.

During the medical day, pupils were again keen to point out that the ambassadors were not like their teachers. A key point was that the ambassadors were 'easy to relate to' and 'easy to chat with', whereas their teachers were not. The reasons were complex and linked to age and the formal positioning of teachers as responsible for instilling and assessing information. One pupil expressed some frustration that she was unable to talk to her teacher 'like a normal person':

Tosin: They are easy to chat with, really.

Jenny: And they're not boring.

Cheryl: Some teachers are boring and it's really hard to have a conversation with them.

John: 'Cause they're just so old.

Clare: Why is it so hard to have a conversation with them and why is it easier?

Jenny: 'Cause they expect us to know everything they know and if I ask, what are you saying, they go on and on and we are just like huh? When they are kind of our age they kind of know roughly what we know.

Tosin: [re. teachers] They don't really relate, so they talk about something and you don't really understand it and then they don't show you, they tell you, but you don't really know.

Cheryl: Sometimes they don't – I was talking to my science teacher yesterday and we were kind of having a conversation and then she just started asking me about particles. It's like – I try

Learning practices and identities

and talk to you like a normal person and then you go and bring our lesson into it.

Medical day

Pupils viewed teachers as didactic and not interested in hearing their views. There were also implicit references to an authoritarian approach by some teachers who 'think they own the place' and 'interrupt' when pupils are talking. This was contrasted to the more negotiated approach taken by ambassadors who 'let you interrupt them':

John: They think they own the place.

Jenny: And I don't like it when, you know, you're talking and the teachers interrupt with something – you're telling them, you're like – oh and by the way – and you don't really say what you need to.

John: They let you interrupt them here.

Medical day

The student ambassadors' discussion of what made the practical activity successful during the medical afternoon provides insights into how the ambassadors were working with the school pupils and reflects observations in my field notes:

Raj: When they got to the practical exercise, which was new to them and didn't have no knowledge about it, they were all receptive, like you're giving them information, and after that you kind of just, like, guided them when you asked them to repeat what they saw and then that was it but yeah, they enjoyed the practical …

Mary: I think it was also 'cause it was something new. Blood pressure and stuff, they probably haven't done it before so it's something that they haven't done before and haven't tried.

Mani: And doesn't involve writing.

Medical afternoon

Pupils were clearly engaged by this experiential learning, by actively trying out medical equipment. As one ambassador comments, the novelty of this opportunity contributes to their engagement: 'It's something they haven't done before.' The pupils themselves were extremely enthusiastic about this activity. Working in a small group, a ratio of three pupils to one ambassador was also described as contributing to the engagement of pupils and providing the opportunity for ambassadors and pupils to talk more informally about their own experiences and plans for the future.

During the medical afternoon, another ambassador identified the benefit of having the opportunity to talk to pupils individually. This small group work was described by ambassadors as allowing for an inclusive and perhaps more negotiated and pupil-led approach, providing all in the group with the opportunity to talk:

> Rachel: I tend to ask, if I'm working with a group, I tend to ask if anybody there is interested in being a doctor and a few will be quite vocal and say, 'yeah', and then somebody will always be, 'yeah', so I'll talk and I'll have them say what they want to say and ask me questions and then try to go to that person separately 'cause they'll be quite timid and they don't want really to show how they feel in front of the whole group so I will talk to them on their own.
>
> Medical afternoon

At several Bankside events, particularly Train Tracks, the summer school and STEM day, and at Royal's medical day and medical afternoon, the ambassadors worked alongside small groups of pupils. They described how they could 'mentor behaviour' by modelling what they expected, by working collaboratively with pupils and by actively engaging in practical tasks. They did not have to manage any difficult behaviour and were involved in making and practice with pupils, where there was no written work or need for assessment. Ambassadors were able to work exclusively with one group, facilitating a calmer working relationship than in school, with little opportunity for frustration from pupils waiting for assistance and attention. Ambassadors were not responsible for the more formal attributes associated with the 'externally determined needs' of the curriculum. This learning environment and their positioning within it enabled ambassadors to develop relationships that were not restricted by the status-driven positionings of pupils and teachers in schools.

Learning contexts also impacted on the pupils' learning about particular subject areas. During both Train Tracks and the medical day, pupils commented on how the real-world contexts had impacted on their thinking about both science and D&T.

During Train Tracks, a group of girls described how working with the student ambassador had facilitated thinking about their 'future':

Learning practices and identities

Ayisha: You can talk about your future.

Train Tracks

The informal attributes of the day's activities were important in this: the activity was student-led, negotiated, and allowed pupils to 'uncover knowledge' which was derived from their own experience. The group explained that the activity had demonstrated to them that a job in design, in an engineering context, provides an opportunity to 'get your ideas out':

Sarah: And this has helped us in deciding what to do in the future … definitely.

Clare: What do you want to do by the way?

Sarah: Something in design.

Clare: How's this helped?

Sarah: In knowing that this is the kind of stuff you're going to be able to do … you can get your ideas out, you don't have to just do the same repetitive thing.

Train Tracks

The day's activities were also identified as helping this group of pupils understand what design in the real world would be like. This group of girls talked quite thoughtfully about what they understood engineering to be, and their account suggests a genuine engagement with and understanding of some of the challenges associated with engineering. This understanding stems directly from the project-based task undertaken during the day:

Sarah: [Engineering] There are some difficult things to consider … I wouldn't say it's an easy subject – it's something where you'd need to use your initiative – you need to put other people into what would go wrong and what would go right.

Ayisha: you need to plan it all out exactly … it's about team work … although it's complicated, you are able to work it out in small stages so you will eventually get there.

Train Tracks

A pupil at the medical day explained how the work they were doing was challenging his thinking about science; that, while it may be 'hard', you 'can actually make something … you're interested in':

> Anwa: I've learned you can actually make something out of it (DNA), something kind of useful; like, something that you're interested in 'cause, like, everyone thinks it's just science and everything's hard, but, like, learning that you can actually make something out of it is quite interesting.
>
> Medical day

This increased awareness of design and engineering and real-world applications of science was facilitated by the ambassadors working collaboratively with groups of pupils during project-based and practical activities. The ambassadors' positioning as subject experts and their work alongside the pupils, negotiating their learning, had evidently contributed to this level of engagement.

Shared learning and subject identities

Pupils discursively constructed ambassadors as learners and students like themselves, working alongside and collaboratively with them in learning contexts with 'experiential' and 'informal' attributes. The impact of the type of activity on the positioning of ambassadors as fellow students was made explicit by one ambassador during the medical afternoon:

> Mani: I think they were able to be more relaxed around us, like because we're obviously students as well and closer to their age and this is pretty much a life environment rather than just being lectured or talked to and yeah, we were just joking and I was pretty sure that they seemed really interested in learning about life, especially the basic life support and they took it upon themselves to test each other as well. It wasn't just us controlling the whole task; it was them getting involved as well and being hands on so that was really good.
>
> Medical afternoon

The age of the ambassadors was mentioned repeatedly by both school pupils and the ambassadors themselves as being significant in this positioning as fellow students. At the medical day one pupil pointed out that even if ambassadors were to become teachers, because they are young they would 'remember what it's like to be a student':

> Janine: I don't really think there's a difference because if they want to become teachers they'll probably be doing the same but they'll not be kind of rude, they are closer to our age than some teachers

and they can understand and realize and remember what it's like to be a student.

Medical day

At the Bankside summer school, Qadira commented on the importance of the age of ambassadors:

> Qadira: Yeah I think … because we're students as well and they're students so they still think okay well they're still students as well so it's actually more – because usually when you've got teachers it's like – oh we've done our schooling ages ago but … we're so close to them it's like – after them – it's us – we're the next step so it's more closer for them to imagine themselves in uni because we are more close in age to them so it's like we're all students together.

Bankside summer school

Qadira identifies perceived similarities in age and status as enabling pupils to see becoming a university student as a logical next step. These discursive constructions of shared student identities appears to position pupils as 'apprentice' university students (Lave and Wenger, 1991) who are learning how to be university students through this interaction.

During the Train Tracks event, one pupil remarked that the ambassadors' status as students was more important than their age in enabling them to 'seem like us'. Rachel, the ambassador she worked closely with, was a mature student in her late twenties and so not particularly close to the pupils in age. Pupils' accounts at these events reflected a strong conception of themselves as learners. Various studies have suggested the importance of biography and established learning identities to how willing and able young people are to engage with learning opportunities (Ball *et al.*, 2000; Brooks, 2003a; Reay *et al.*, 2009; Crozier *et al.*, 2010; Evans *et al.*, 2006). Existing research into ambassador work has identified a lack of interest in HE among pupils as obstructing relationships and even contributing to a sense of social difference (HEFCE, 2010a; Taylor, 2008; Gartland, 2014). The pupils contributing to this research, with the exception of a few Year 8 pupils attending the medical day, were all positively orientated to university. At the medical day, though from low-achieving schools, all pupils attending had been identified as G&T. Pupils were aware that their success as learners had led to their being selected to attend the event and identified as different to other pupils in their year:

> Anwa: Miss said that we … she said we were very special, like, to be there.
>
> Medical day

A group of boys at the medical day discursively constructed ambassadors as role models. Their shared youthful student identities and learning identities contributed to this:

> Ton: They are not old but they are young, so they are basically like us, but, like, in a few years' time.
>
> Dixon: I think they understand us more than the older people … because everyone is kind of from the same generation – kind of even though they are an age apart – some ages apart but like …
>
> Ebo: They show you what they've done.
>
> Ton: Many of them might have just started university.
>
> Ebo: They are like a role model.
>
> Dixon: So they know what it's been like, to be like a child and to be like us and, like, I think they all like to learn and we like to learn as well.
>
> Medical day

Other pupils at the medical day reiterated this discursive construction of ambassadors, explaining 'they're learning and we're learning as well'. Again, this shared learner identity contributed to a sense of proximity and to pupils' view that ambassadors could relate to them: 'they're learning, they speak the same kind of language we speak':

> Tosin: Maybe they are learning as well – learning from other people's points of view to what things are and how they understand them, and they are taking it all in, so they can teach other people different points of view and how other people feel about whatever topic and stuff …
>
> Jenny: You know, for, like, different stuff they are in a more higher education than we are – we are, like, in secondary and they are in uni, so they obviously know more than us – they try to explain it to us like how they would like it to be explained to them.
>
> John: They're not teachers, so they're learning from their university teachers. They're learning as well and we're learning as well … .

Learning practices and identities

Jenny: They're learning, they speak the same kind of language we speak.

Medical day

At the Bankside summer school, one pupil also described the ambassadors as academically 'smarter' than himself, explaining that the ambassadors' status as university students, as academically 'smarter', is the reason that the pupils are prepared to 'listen so well'. He implies that such academic achievements should be respected:

Michael: I think the reason why we probably listen to them so well is because they study here – they study engineering here, so we know they are much smarter than us.

Bankside summer school

Michael positions the ambassadors as experts, reflecting his own positive orientation to academia.

The subject identities of ambassadors and pupils were an important aspect of their shared learner identities. For the group of pupils spoken to during Train Tracks, the subject focus of the activities and the expertise of the student ambassador were important, as these related to their own interests. This was illustrated by the enthusiastic response to the student ambassador of one of the Year 10 girls:

Ayisha: I always wanted to do further education within a certain field, but knowing how someone else went through the same thing – yeah, kind of drives you more.

Train Tracks

This was also the case for pupils attending the G&T summer school at Royal. During a focus group, one pupil identified that the pupils and ambassadors all 'have very similar ideas' and that this is rooted in the fact that they all share an interest in medicine:

Martin: Yeah, just really friendly and really relaxed and (inaudible) … about everything. Obviously, most of us here want to go into medicine – so we have very similar ideas.

Kate: They're just so, like, nice – they're just really easy to talk to and the thing that we need.

G&T summer school

At the medical day, the expertise the ambassadors demonstrated during activities clearly impressed pupils. There was some suggestion from one pupil's account that he was even positioning himself as a potential university student, or even a Royal student, through his reference to 'what you would learn when you are here':

> Dixon: And one of the students – they were talking about DNA at the start and it shows you how much they learn 'cause they knew lots of stuff, so it basically shows us, like, what you would learn when you are here.
>
> Medical day

As Liang and Grossman (2007) discuss in the context of mentoring relationships, while there is much emphasis on matching backgrounds in terms of gender and ethnic identity, similarity may be indicated by other qualities such as shared interests and geographic proximity. The discovery of shared subject interests by the pupils during interactions while undertaking practical activities certainly appeared to contribute to identification between school pupils and ambassadors.

What are pupils learning?

Student ambassadors' accounts frequently included constructions of themselves as sources of information, advice and guidance. Practices of exchanging information were closely linked to how ambassadors were positioned by the learning context and processes. When positioned as authority figures or given more formal roles, such exchanges were less likely. In small groups during project-based and practical activities, these exchanges occurred and ambassadors were seen as trusted, hot sources of information, as outlined in Chapter 5.

As sources of information, advice and guidance, however, ambassadors were limited by their positioning within marketing discourses as promoters of both university and particular subjects, constraining and shaping the advice that they provided. Ball *et al.* (2000: 10) point out that there is 'a general tension or confusion in the education market between information-giving and impression management and promotion'. This tension was evident in the aspiration-raising work of student ambassadors. Ambassadors were genuinely keen to help and advise pupils, but they were equally keen to encourage them to pursue engineering or medicine. The reliance by ambassadors on their own narratives, their personal histories, in their advice to pupils also contributes to 'opacity rather than transparency' about HE (Ball *et al.*, 2000: 10).

Providing such individualized information to pupils, particularly about medical degrees, appears problematic. Ambassadors' own narratives do not provide pupils with a transparent view of how to access HE courses in medicine.

Learning practices and identities

School pupils are positioned by activities and through these narratives as being fellow learners, like ambassadors, and as such, as potential medical students at Royal. While this may be a likely positioning for some pupils on the G&T summer school, the extended degree programme only offers a handful of places to pupils who do not achieve the prerequisite top grades. The individualized and isolated success stories of the student ambassadors on this programme ignore the structural barriers the majority of these pupils face. As Reay *et al.* (2005: 161) suggest, the 'combination and interplay of individual, familial and institutional factors' serve to constrain the choices of state-school pupils from ethnic and poorer backgrounds. It may be that, by encouraging these pupils to aspire to study medicine, ambassadors are encouraging them to pursue a dream that will prove unattainable. As Delgado's (1991) article powerfully points out, there is something irresponsible and even unethical about encouraging young people from deprived areas to aspire to professions that in reality are virtually closed to them.

Conversely the 'comfortable' experience of pupils on the summer school and their identification with ambassadors at Bankside may actually constrain rather than raise pupils' aspirations, an issue raised in Chapter 5. A desire to find an institution where he felt 'comfortable' was the reason given by one student in Reay *et al.*'s (2005: 92) study for turning down a place at an elite university. As the authors suggest, pupils' 'choice of university is … a matter of taste and lifestyle in which social class is a key determinant of choice' (ibid.).

The data also reveal interesting patterns in the learning about STEM subjects during activities. Shared learning and subject identities between ambassadors and pupils were clearly important to this learning. Ambassadors' work with pupils during the maths workshops was defined by credentialist discourses, dominant in schools who are key stakeholders in WP work. Ambassadors' interactions with pupils in these contexts focused exclusively on practice for GCSE maths exams and there was little identification between ambassadors and pupils during workshops, as ambassadors were clearly positioned by the formal attributes of the learning environment as teachers of the maths syllabus. However, as Williams *et al.* (2010) discuss, this focus on the 'exchange value' of maths encourages identities among pupils as 'surface learners' rather than as 'users of mathematics'. Ambassadors may serve to support this surface learning, but pupils and teachers were both dubious of the quality of the support provided.

In contrast, Train Tracks and the medical day challenged and extended pupils' understanding of science, engineering and design and technology (D&T) by providing pupils with the opportunity to engage in practical activities that provided insights into real-world applications for these subjects. Working alongside ambassadors in small groups facilitated discussions about the task as well as about STEM subjects and the ambassadors' own experiences and trajectories. The location of the activities at a workplace and at a hospital may be significant to pupils' engagement and to this understanding. These informal attributes of the learning contexts appeared significant to the learning that was taking place. However, the engineering camp illustrates that practical activities do not necessarily engage pupils in this way. The positioning of ambassadors as supervisors responsible for the behaviour of pupils and the didactic approaches taken by supervisors in relation to pupils' work undermined relationships. Conversely, the lack of structure and timings for activities during this event led to pupils losing focus and engagement.

Where ambassadors worked collaboratively alongside pupils on practical activities, pupils viewed them as both 'like themselves' and the 'next step'. Drawing on Lave and Wenger (1991), these pupils can be conceptualized as 'apprentice' university students. To become an expert member of the 'community of practice' of university students to which they aspire, they acquire knowledge from the ambassadors informally and unintentionally through 'situated learning'. Post-structuralist theories about creating individual subjectivity and being embedded within particular cultures and times are also useful. According to Butler (2004: 45), people are essentially social and 'are comported towards a "you"'. Butler's concept of performativity resonates with student ambassadors' conscious performance of identities as perfect maths, engineering or medical students, with school pupils either joining in or performing in opposition to this. It may be that pupils' future identities as university and STEM students may be constituted through such performances.

Chapter 7
Social relationships and identities

The idea that adults working with young people are role models is a ubiquitous one. So how do student ambassadors respond to this dominant discourse and how do they actually practise being role models? Is this notion useful in developing ambassador and mentoring work or does it operate to obscure more complex processes of dis/identification that actually take place between pupils and young adults?

In the previous chapter I considered the importance of learning contexts and learning and subject identities to processes of dis/identification. As the widespread practices of matching backgrounds in mentoring would imply, other aspects of pupils' and ambassadors' identities are also significant to these processes. Learning and subject identities 'intersect' (Crenshaw 1989; Mirza, 2008; Morley, 2012) with pupils' and ambassadors' gendered, ethnic and cultural identities, either facilitating or constraining how they identify. In this chapter I consider pupils' discursive constructions of ambassadors as being like friends, relatives and even role models and the significance of the social relationships that can develop between ambassadors and pupils in some learning contexts.

Role models?
Accounts of ambassadors' work conformed to WP discourses of individualization and aspiration-raising, and the ambassadors were seen as aspirational role models for pupils. Organizers' descriptions of ambassadors and ambassadors' descriptions of themselves frequently included their being role models. These accounts appeared to reinforce ambassadors' positioning as marketing their HEIs and subject areas, as the only obvious way of enacting being an aspirational figure is to encourage others to copy what you have done. However, within the discursive constructions of ambassadors as 'role models', there were also discourses relating to ambassadors' identities and to teaching and learning.

Role models and marketing
During training at Royal, ambassadors were explicitly told that they were role models:

> You don't have to say everything's wonderful – be honest, be yourself – it's not a marketing deal but you are very much a role model.
>
> WP coordinator, Royal

This coordinator went on to tell ambassadors that 'you're exciting anyway because you're students – you're cool'. This statement holds the assumption that pupils will admire and aspire to emulate the 'cool' ambassadors.

At Bankside there were similar discursive constructions of the ambassadors. During the training of engineering ambassadors, the fieldworker told them to 'act as a role model, share your experiences', which again implies the positioning of ambassadors as aspirational figures. During the meeting at Bankside about ambassadors, the head of school outreach commented:

> If they [ambassadors] are the best of Bankside, you would hope they would provide an example of the sort of behaviour younger people would want to emulate.
>
> Head of outreach, Bankside

This positions ambassadors as role models and carries assumptions that ambassadors acting as role models – 'the best of Bankside' – will effectively market their institutions. Ambassadors, then, are taught that they are aspirational figures for school pupils and have to enact being a role model in the different learning contexts in which they are placed. This enactment inevitably involves promoting their university, course and subject as their student status is central to their being role models. However, other aspects of ambassadors' identities also seem important in their enactment as role model.

Ambassadors' discursive constructions of themselves as aspirational role models for school pupils featured heavily in conversations. Unsurprisingly, being a role model is interpreted by Adam as successfully marketing his own course:

> Adam: They see us as a role model – they want to do it – like two of the students actually told me that. I say, 'what I do is I do design and I have to tell my lecturers what I do and why I'm doing it', and they are like – 'yeah, I want to be an architect'.
>
> Bankside summer school

The extent to which ambassadors' accounts of their status as role model are imbued with familiar marketing and other dominant discourses was clear in discussions among Royal student ambassadors at the medical day:

> Chanelle: In some sense maybe you feel like a role model; not that you're around them all the time, but it gives them – what you're doing is an inspiration to them and you are also an adviser to them.
>
> Falak: Like a mentor sort of thing.
>
> Simi: It's just …
>
> Chanelle: Sometimes that's all you need though, if you're honest, to get inspired.
>
> Adi: Tony Blair's role as Middle East emissary like – we going spreading the good word of medicine and really to kids who might not be thinking of it [laughter].
>
> Medical day

Inspirational role model and 'advisor' are used interchangeably in Chanelle's account and it is clear that the 'advice' that these ambassadors are keen to provide school pupils with is intricately tied up with their desire to promote the study of medicine and the MAS among these pupils.

The religious references drawn on in Adi's explanation, that it's like 'spreading the good word of medicine', presents his work as an ambassador as being akin to the work of an evangelist, converting non-believers to believe that they too can study medicine. In a later description of a visit he made to a primary school, Adi refers to his ability as an ambassador to 'inspire the pupils'. Drawing on similarly religious frames of reference, he constructs himself as 'sort of a higher being' in the eyes of school pupils:

> Adi: I went into a primary school during term and I was doing a cardiovascular day and the way the kids looked up to me and I wasn't really doing all that much, I was just … showing them which way round to walk around, things like that, but they see you as sort of a higher being. To know that you can inspire someone just by talking to them and just by listening to them even is a really nice thing to happen.
>
> Medical day

Adi's positioning as role model promoting medicine correspondingly positions pupils themselves within discourses of 'deficit' as the unenlightened that need to raise their aspirations.

Modelling behaviour

The comment made by the head of outreach at Bankside, that ambassadors 'would provide an example of the sort of behaviour young people would want to emulate', highlights another meaning of role model. Here ambassador work was seen as providing a model of behaviour for pupils to emulate and has a disciplinary function. This modelling was seen as a way of influencing and even controlling the behaviour of school pupils.

Ambassador training at Royal briefly outlined that ambassadors should model the behaviour that is hoped for and expected of school pupils in terms of engagement with tasks and activities: 'Enjoy yourself – if you're enjoying what you're doing – they'll enjoy themselves as well' (WP coordinator, Royal). Bankside ambassadors reported that they were explicitly instructed by organizers to provide pupils with models of behaviour as a way of ensuring their engagement with activities and encouraging positive behaviour.

This interpretation of role model was also evident in ambassadors' accounts, particularly among engineering ambassadors at the summer school. Ambassadors described how they model behaviour they hope for from pupils to encourage their participation. Adam explained how his involvement in the afternoon rap session enabled pupils to overcome their self-consciousness and participate:

> Adam: We just encourage them. Because the DJ rap session was, like, two days ago in the afternoon session, they didn't want to do that – they didn't even want to do it because they had things written down that they couldn't say and I was like – 'okay, fine if you don't want to say it, I'll say it for you', and they got encouraged.
>
> Bankside summer school

Qadira described how pupils 'copy everything we do' and that they had been told during training to 'pretend to like' what they are doing. It was apparent from her account that she found this constant display of enthusiasm quite difficult to maintain: 'We have to be on our best behaviour. It's quite hard for us.' Qadira again stressed the importance of engagement in activities by the ambassadors as a way of encouraging pupils:

> Qadira: … and getting involved with the activities that they do as well, so they can be encouraged at the same time, 'cause if they don't see us doing it then they'll be a bit discouraged.
>
> Bankside summer school

At times, ambassadors' accounts of marketing university and subject areas and modelling ideal student behaviour converged. Adam explained that modelling appropriate behaviour facilitates his own position as a role model:

Social relationships and identities

Adam: In summary I can see that they actually want to get to uni because they can see me do it. We are just trying to be like them – trying to relate to them, trying to get them to do things and they've got this *pilot* – like, 'I really want to go to uni. I want to be like these guys'.

Bankside summer school

Qadira explained how participating in activities alongside pupils enables pupils to see that 'engineering is not that bad':

Qadira: We assisted in the activities, as in we partook ourselves, which was really fun. Yeah, I think when they see us having fun doing it they'll be like – I think – you know, that engineering is actually not that bad.

Bankside summer school

During the engineering camp, another ambassador described her work as 'showing' pupils what studying engineering is like, which she achieved through working with them on an engineering project during the camp:

Akua: Showing them what it's going to be like – it's not different from any other course. Some of them are saying it's hard – it's showing them it can be fun – find a solution, not just lots of reading … it's not as difficult as people make it out to be … it's like building the plane … bit by bit starting and finishing – you know where you are going – you have an end product which is always achievable.

Engineering camp

Like Qadira on the summer school, Akua views her work as modelling, through her own behaviour, what is expected of pupils. She believes this will challenge their perceptions of engineering as 'difficult' and enable them to see themselves as potential engineers.

The accounts of ambassador work as being to model behaviour and enact the ideal student has been a useful lens through which to consider pedagogy and ambassador–pupil relations. The data collected illustrate how these enactments and pupils joining ambassadors in this performed identity can be powerful in providing pupils with opportunities to engage in possible and 'viable' new 'ways of being' (Davies, 2006). These joint performances can be considered as performative according to Butler's theories: 'the subject who has been named is able to name and make another' (Youdell, 2003: 6). By participating in such a performance, pupils are in a process of 'subjection':

> Subjection is, literally, the *making* of a subject, the principle of regulation according to which a subject is formulated or produced. Such subjection is a kind of power that not only unilaterally *acts on* a given individual as a form of domination, but also activates or forms the subject. Hence, subjection is neither simply the domination of a subject nor its production, but designates a certain kind of restriction in production.
>
> Butler, 1997b: 83–4

The potential impact of these shared performances was illustrated in the accounts of ambassadors who had participated in activities when they were school pupils themselves. One student explained how being involved in the summer school at Royal had made her 'look up to' the ambassadors and consequently made her want to become an ambassador herself:

> Vanessa: Yeah, at Royal, and I recognized one of the ambassadors; I don't think she recognized me when I came back. I said hi and she sort of said, 'I don't know who you are' [laughter].
>
> Clare: Did you remember the ambassadors?
>
> Munira: I don't remember her very well but I remember that I made her skip lunch so she could go to the … meeting with me. I think she might remember me but also I made a few friends on the actual summer school and they applied for the same course and they're in my year and we just know each other – it's like, oh hi – you know, it's quite cool and I told him about how I'm doing the summer school now and he's like, 'really, why are you doing it' – and he's like, 'why didn't you tell me about this', and when you attend these schemes you actually want to become the student ambassador 'cause I think you look up to them then after you get more involved.
>
> Vanessa: Yeah, definitely.
>
> Than: It's a nice circle.
>
> G&T summer school

Ambassadors also discussed how relationships they establish with pupils while working with them during visits facilitates the building of relationships if pupils then also progress to the university. Alicia describes how it 'makes them feel more comfortable' and Carla, who says she is not a 'typical' Royal student in terms of class and race, worked with ambassadors from Royal herself when she was at school and explains that 'it can be quite scary coming to university' and having the security of having 'someone you know ... is quite nice':

> Carla: So you had to basically know all the physics students who come to Royal – I just see them all the time, there's not that many of them and it's great when they actually come to Royal and you see them in the first year and you talk to them, build a relationship with them ...
>
> Alicia: It makes them feel more comfortable.
>
> Carla: Mm, because it can be quite scary coming to university, but if you feel you already have someone you know, who is quite happy to ask questions, it's quite nice.
>
> G&T summer school

These ambassadors' accounts suggest that feeling comfortable or at ease was partly achieved in their own and others progression to Royal, despite their backgrounds, through social relationships formed with ambassadors and fellow pupils during outreach activities.

Cultural locations and intersecting identities

A question repeatedly posed in the mentoring literature is whether matching the backgrounds of mentors and pupils is important (Klaw and Rhodes, 1995; Sanchez and Colon, 2005; Cavell *et al.*, 2002; Jackson *et al.*, 1996; Kalbfleish and Davies, 1991; Sanchez and Reyes, 1999; Chen *et al.*, 2003; Ensher and Murphy, 1997). An important related question is, what aspects of identities are significant? The need for female role models and mentors in STEM has been much discussed (Siann and Callaghan, 2001). A more germane conceptualization for these young people may be the notion of multiple identities. Mirza (2009: 80) draws on Crenshaw's concept of intersectionality and identifies the need to understand 'the value of an intersectional analysis that reveals the multiple identities' of young people.

The data reveal that a combination of gendered, raced, classed, cultural, youth, student and learning identities, as well as orientation to particular subject areas all contributed to pupils' relationships with student ambassadors. Pupils 'actively negotiate subject positions within discursive constraints' (Hughes, 2001: 276). As Liang and Grossman (2007: 251) discuss, in mentoring relationships, there appear to be 'complex interactions between demographic characteristics and multiple aspects of a youth's identity'.

The Excellence Challenge Interim Report (HEFCE, 2005) outlined how university students working with school pupils can 'play a part in breaking down cultural barriers' and make higher education 'cool in the schools'. In most of the learning contexts considered in this study, pupils were already positively orientated to university, and pupils and ambassadors were generally quite well 'matched' in terms of their backgrounds.

There were several female ambassadors at each engineering activity and in all contexts male and female pupils had the opportunity to interact with someone of the same gender. Pupils and ambassadors were from local south east London schools, with a large number sharing a Black African heritage. The G&T summer school at Royal was the exception, comprising pupils who appeared to be predominantly middle class, drawn from comprehensive schools across London.

Matching aspects of identity

During the careers session Ed described how he felt that his male identity and mixed heritage enabled male Indian pupils to feel 'more relaxed':

> Ed: I'm not really black. I'm half Indian … it makes them more relaxed – the Indian ones when they find I'm half Indian – one boy today – he changed a little bit – he relaxed more.
>
> STEM day

This view was reflected in observations where I noted that a small group of Asian boys appeared shy and reluctant to contribute, but were actively engaged and talkative during the car-making exercise when working with Ed.

Shared gender and ethnic identities could also help explain the speed with which one student ambassador, during Train Tracks, was able to develop close relationships with a group of Year 10 girls. Both the ambassador and the girls were from Asian backgrounds. This, combined with their shared subject interests, contributed to Rachel's apparent positioning by pupils as someone that they aspired to be like. This group of Year 10 Asian girls talked explicitly about how they saw the ambassadors as aspirational figures. One student explained that 'by seeing them it inspired me to learn more'. Others in the group explained how this motivated and inspired them:

> Ayisha: I always wanted to do further education within a certain field, but knowing how someone else went through the same thing, yeah, kind of drives you more.

Social relationships and identities

Sarah: ... they were inspiring to me because they told us about what they do.

Train Tracks

Rachel had acquired this positioning as an aspirational figure in pupils' eyes through narrating her own experience of making subject choices. She enacted being a role model by talking enthusiastically about her own university course and trajectory. Rachel was aware of the students' enthusiasm, describing how their 'eyes were wide open talking about it'.

Another ambassador described how ambassadors' racial and gender identities and their 'backgrounds' enable school pupils to see a 'reflection of themselves'. She suggested that this can lead to a 'big response' from pupils who share these aspects of the ambassadors' identities:

Remi: I do think if a black male is speaking to a young black boy then they see a reflection of themselves in that person and, although it might not be obvious, it does mean a lot – if you see – like, you think, you see, 'I came from the same background, I can do this', because usually they're so used to seeing the same kind of doctor, the same kind of – so it does have an impact. I find when I speak to the guys, I don't get no response – they're usually like – that's it, but when I speak to the girls I get quite a big response, but I do think race does go into it as well. I think it's a big part of it.

Dan: I didn't think for Year 8s – maybe for more, like, Year 10s.

Remi: I think you'd be surprised, because sometimes they've never seen a young, black, professional male or a black male who is in a position of, like, say, university or doing something other than the typical, the stereotype, so it's an underlying subliminal message but it does mean something to them.

Medical afternoon

An ambassador at the Bankside summer school commented on the immediate visual impact of seeing female and minority-ethnic ambassadors:

Qadira: Yeah, I think it [race and gender] is really important and I think it must be the first thing they must pick up on when they meet us for the first time. They must look at us all together and think, 'okay'.

Bankside summer school

123

Qadira talked specifically about how this is different in different geographic locations and how, in south London, being black 'would help' the 'black youngsters', but that there is a need for a mix of ethnicities so that the pupils can see that 'anyone can go into it'.

The gender of the ambassadors was seen as particularly significant in the context of engineering activities. Accounts of both ambassadors and organizers reveal the challenge posed for pupils by female engineering ambassadors. According to Freya, a young woman studying petroleum engineering,[1] pupils responded to her with surprise combined with admiration:

> Freya: People don't know about it – petroleum engineering – people think I'm a girl so I'd be doing something different – people think I'm like Einstein – but I'm not, it's just a normal course.
>
> STEM day

A response also described by Qadira at the summer school, who explained how one female pupil had even asked if she was in the 'right room' at the start of the engineering course:

> Qadira: A lot of the girls say – 'oh, so what engineering do you do?' Always, always, always, always – because they always have it in their mind, ah, that it's just guys. Being a girl there, it always seems to get them, like, interested. Immediately they always think to themselves – on Monday somebody asked me – 'are you meant to be in this room?' I was like – 'yes'.
>
> Bankside summer school

Ambassadors' accounts suggested that different aspects of their identities affected the relationships they developed with pupils. Carla explained how she thought her race and gender contribute to her being viewed as a role model:

> Carla: Yeah, I think if, like, I went to, 'cause I speak to a lot of black girls who don't like science – black typical, that bad view of science, hopefully when I speak to them I say that normal people can do physics and science and I'm still not like a total geek. Hopefully I am a good role model because of my colour and where I am from – a single parent family – I hope it just kind of – 'cause I do talk to them about stuff like that – I hope that, because I am from that situation, it might hopefully inspire them to go on to do it not just be, I don't know, think they can't achieve.
>
> G&T summer school

Social relationships and identities

Carla appears aware that her ethnicity and gender intersect with her social class – 'where I am from'. She explained that she talks to pupils 'about her 'single-parent family' background and that she hoped that, 'because' she 'is from that situation, it might inspire them'. There is a discourse of difference both in terms of racial identity and also of classed identity. Though class is not discussed directly, ambassadors' accounts indicated an awareness of their own and the classed identities of pupils and how, at times, these contrast with the more dominant middle-class identities of students on medical degree programmes at the university.

There is an implicit reference to the classed and ethnic differences implied by the geographic locations of pupils and students in a comment made by Carla during the G&T summer school at Royal, when she suggests that west London is 'a different world' to south London:

> Munira: Yeah, I think so – in terms of the kids, we often work with kids who are from south London; I'm not from south London. I do find it a little bit more like I have to sort of find a common ground which is not kind of background, racial or kind of area-based and it tends to be like, music or something that I can sort of wing.
>
> Clare: Where were you?
>
> Munira: West London – not far but it's different.
>
> Carla: It's a different world.
>
> G&T summer school

Coming from 'the same area' is one aspect of the ambassadors' identities that they suggest is useful in providing a link with pupils. Their 'race', 'religious beliefs' and 'colour' are also mentioned as being potentially useful in enabling pupils to 'associate with' the ambassadors:

> Munira: … 'cause like this summer school, it doesn't really matter who you were, but sometimes if the children are kind of withdrawn they turn to someone who they can associate with, so – same colour, same race or the same kind of belief, religious beliefs even.
>
> Carla: Or even area – if you live in the same area.
>
> Munira: Oh yeah, area – we're from the same area.
>
> G&T summer school

Two of the medical day ambassadors attended the same schools as two groups of pupils at the event and they explained how this contributed to a shared sense of identity:

> Simi: Usually it's things like – 'no you didn't – do you know Mr. Clegg – yeah, I loved that teacher' [laughter].
>
> Adi: It's usually teachers – yeah.
>
> Chanelle: And then for some reason you go away from the hall, because before you have to speak quite proper and when you start to find yourself – you speak the same lingo.
>
> Adi: On common ground.
>
> Simi: And yeah, they speak more openly – like, 'no, I actually do want to go to uni, but …' because they can relate to you; instead of you becoming Miss you become someone who is just like them, 'cause for a while they kept calling me Miss – like, 'I'm not Miss'.
>
> Chanelle: I ignored a couple 'cause they started calling me Miss.
>
> Clare: So they stopped calling you Miss …?
>
> Chanelle: Yeah, they do because they think you're one of them, like.
> Medical day

Chanelle identifies this shared school background as contributing to pupils and ambassadors shifting their style of speech to 'the same lingo' and to pupils being more open about their plans for the future. It is also worth noting that Chanelle identifies this as shifting the positioning of ambassadors away from that of 'teachers': in the pupils' eyes they are no longer 'Miss'.

Pupils also identified the 'lingo' that ambassadors used during the medical day, enabling them to understand and relate to them:

> Tosin: When they talk freely they don't talk really, really formal like.
>
> Abi: If they talk like your teachers … we might not understand something because they're, like, over and over again, but like they not know – it's probably the first time we've heard about it – like, they break it down for us and they don't talk, like, all confusing – they speak how we do, kind of.
>
> Tosin: And they are more straightforward – you know teachers, they just go on and on …

> John: I heard one of them say, 'butters' [laughter] – slang.
>
> Medical day

The reference to the term 'butters' suggests that part of this shared lingo is a shared knowledge of slang widely used by teenagers in south east London and closely connected to the dominant hip-hop culture among black teenagers and particularly black males (Kitwana, 2002; Reese, 2004; Gosa and Young, 2006).

Ambassadors' and pupils' accounts reveal that the practice of matching backgrounds is significant to how pupils and ambassadors identify with each other. Intersecting identities including ethnicity, gender, speech style, geographic areas and which school young people attended, all appear to play a part in this process. This is perhaps unsurprising and is already acted on in these institutions, with organizers attempting to match pupils and ambassadors as closely as possible. However, it is arguable that this practice of matching to encourage progression, far from 'breaking down cultural barriers', works to support them. At Royal and Bankside ambassadors can be seen to be supporting the status quo in terms of patterns of participation within a highly stratified HE system: ethnically diverse and working class at Bankside and middle-class G&T pupils at Royal. I have also discussed previously (Chapter 5) how ambassadors' accounts of their own individualized success is problematic, correspondingly attributing failure among pupils to progress along a similar trajectory to their lacking the appropriate individual qualities rather than being bound to their classed, ethnic and gendered positioning (Ball *et al.*, 2000; Evans, 2007; Burke, 2012).

Dis/identification – 'She's a vegetarian, but we like Kentucky chicken'

During the maths workshops, pupils talked in some detail about how ambassadors were similar or different to themselves. These discussions reveal pupils' intersecting identities and processes of dis/identification with ambassadors. Perceived differences between themselves and ambassadors were key to this dis/identification, and gendered identities played a central part in this process:

> Yvonne: Both – similar and different …
>
> Bim: That man was different because like race and stuff. She's from India and I'm from Africa so … and, like, the way she will dress will be different from the way I would dress. I hope her type of music would be different from my type of music, 'cause she is into Indian stuff like that and I listen to hip hop and stuff.

Yvette: Did she tell you?

Bim: Yeah – she was playing Indian music when I was walking, but we are similar in a way because – I don't know – we're not similar.

…

Bim: And you know Helen, she's, like, up to date and stuff, so that is how we are similar. She's very energetic and fun and bubbly and that's kind of like us.

…

Bim: They all look too smart. When they come in they're really big bags – they've got their big shoes and they look too smart.

Janine: Big shoes [laughter].

Clare: What does that tell you about them?

Janine: They're educated.

Yvette: It looks like they're people that get bullied – yeah [laughter].

Bim: Like they're sweet.

Leticia: Yeah.

Yvette: I think about that – anyone that's got a big bag and flat shoes – I think their life must be so boring [ouch – laughter] I do, you know … I don't know what they have in their bag but …

Leticia: It's so big like a rucksack.

Yvette: And it's always packed as well. I think half of the things in there they don't actually need – that's what I think.

Clare: Okay, so they've got a big bag and that makes them different because you wouldn't carry a big bag.

All: No.

Bim: Not a rucksack, I wouldn't carry a rucksack

Maths workshops

The first difference that Bim identified is racial; the ambassador she refers to is Caribbean, while her heritage is African. She also identifies another ambassador as being different because she is Asian. However, her explanation of this difference quickly moves into aspects of youth culture with contrasting taste in music – 'I listen to hip hop' – being central.

Bim then tries to think of similarities, but comes to the conclusion 'we are not similar'. While they talk with some warmth about some ambassadors – Amir is described as 'humble and calm' – it is evident from their descriptions of these ambassadors' 'big shoes' and 'big bags' that they see them as very different to themselves, again particularly in terms of style and youth culture. While the pupils view the ambassadors as 'educated', they are also described as 'people that get bullied' and certainly are not aspirational figures for these young women. Indeed, these accounts resonate with Hey's descriptions of the girls in her study where she found peer groups identifying with their own groups and having oppositional stances to 'others':

> Here we have pupil subjects who belong to 'our group' as opposed to 'the other' to specific classes and races; to sexualities that are proper as opposed to being the other side of the hill.
>
> Hey, 1997: 138

The male students in these accounts appear emasculated and infantilized: they are 'sweet', harmless, in danger of being bullied, so in need of being looked after. The conversation moved on to focus on Adam. While, at 30, Adam was considerably older than the other student ambassadors working in the maths workshops, he was viewed by this group of girls as desirable and contrasted to the other ambassadors in his style, though he was not admired for his contribution to their maths work. Adam shared a Nigerian heritage with several of these girls. For this group of girls, as for the black girls in Youdell's study (2003) the focus for 'proper' sexual desire is intimately related to both racial and cultural identities:

> Bim: Now, Adam can dress.
>
> Janine: I remember when he was some red top – and he looked nice, I'm not going to lie. When he comes to clothing he's on points. But maths – that's a different story. He's a nice view – a nice sight.
>
> Maths workshops

The conversation then moved back to the other ambassadors who were again discursively constructed in negative terms as doing 'maths all the time' and not having a social life. These accounts resonate with the hostility to 'boffin' identities expressed by the working-class white girls in Hey's study (1997). As earlier in the conversation Bim had described Helen, a student from another elite HEI in London, as 'bubbly', 'fun' and 'up to date', 'like us', I asked about her specifically. While there was some admiration for her being 'smart' and not a 'dumb blonde', despite her appearance, it was very clear from their discursive constructions of Helen that none of the girls identified closely with her. Indeed they quickly progressed to *imagined* aspects of her identity, such as that she is a vegetarian who likes organic food, to explain how she is different to themselves. Yvette even compared her to 'people in America', which metaphorically illustrates her sense of distance:

Yvette: Similar interests?

Bim: Oh, way different.

Leticia: Maths all the time.

Bim: And I don't think they socialize or anything like that.

Clare: Even Helen?

Bim: No, Helen – yeah, but not the rest.

Janine: Most people talk about blonde girls being dumb girls, but she is, like, smart.

Bim: Yeah, she goes to clubs and then she works – she's one of dem clubbin', running girls.

Clare: She looks healthy does she? She doesn't look like …

Janine: She looks like one of dem girls that will talk about 'I don't eat animals – I eat leaves' I think she's one of them type of girls.

Laticia: She looks like a Miss P type. She'll talk about – 'I don't eat animals, I eat miso' – but she looks smart.

Clare: But she doesn't eat meat?

Janine: I don't know – she looks like one of dem girls that will speak to me about things that I don't even care – like, 'I don't eat meat, I don't do …'.

Yvette: No, she doesn't talk like that. She's like, you know those people in America – in the countryside and there's like horses and cowboys and stuff from Alabama – that is what she's like.

Clare: Except she's not American, is she – she's English – do you means she's countryside English?

Yvette: Organic people – the ones that kill their own chickens.

Janine: Yeah, something like that.

Bim: I saw her jogging.

Leticia: [re. Miss P] Every time I am walking with a pack of chips, she's – 'let me see that' – the protein or colour – I am like, 'whatever – I want my chips'.

Bim: She's a vegetarian, but we like Kentucky chicken.

Maths workshops

These pupils consciously constructed themselves as different. Their gendered identities were central to this difference, resonating again with the girls in Hey (1997: 131) who 'constructed identifications against as opposed to with other girls'. In this assertion of difference there is also an exuberant defiance of the 'respectability' associated with being white and middle class like Helen and Miss P. As Taylor (2008: 159) explains in her study of ambassadors, these pupils are 'very much the ones "on top"' in this classroom context. They reject Helen's healthy active lifestyle – 'I saw her jogging' – and embrace instead 'Kentucky chicken' and their own youth culture. This youth culture is distinct in its urban and ethnic identity and style, food and ways of being. So a slim, white, blonde girl at an elite university who *probably* eats healthy food, appears very unlikely to become an aspirational role model for these girls, despite her social skills, aptitude in maths and good looks. If these interactions are seen as performative, they are constituting these pupils as inappropriate and even in opposition to the university and subject identities that ambassadors represent. These interactions are likely to reinforce rather than challenge these pupils' understanding that identities as students of mathematics and as students at elite institutions are not 'viable ways of being' for people like them (Davies, 2006: 430).

Discourses of difference and dis/identification were also present in the accounts of ambassadors. During the summer school, ambassadors compared the pupils at the events with pupils targeted by Aimhigher at the Easter School earlier in the year. Ambassadors all described the Aimhigher pupils as being more challenging. Marvin explained that 'the kids weren't terrible … they were just different'. Carla went on to explain that at meal times 'it was like a rush for the food – take as much as you can and moan about it'. The ambassadors attributed the difference between these groups of pupils to the G&T status of those at the summer school:

Clare: There is quite a big difference, is there?

Arla: Yeah, 'cause they were happy to work on their own here, but not ...

Than: It wasn't gifted and talented, was it?

G&T summer school

Carla referred to the pupils on the Easter course as 'a bit rough'. This account resonates with Taylor's (2008: 155) findings 'that social class is mobilized in constructions of the "good student" as against the "bad pupil"'. As Skeggs (1997: 3) identifies, 'the classification by and of the working classes into rough and respectable has a long history' and 'respectability' is in many ways synonymous with middle-class identities:

> The working classes are still 'massified' and marked as others in academic and popular representation where they appear as pathological.
>
> Skeggs, 1997: 3

There was an awareness among the ambassadors that there are potential Royal students among the G&T pupils at the summer school and, conversely, that many Aimhigher pupils were not potential Royal students. Carla, who indicates that she is not from a privileged background but from 'a single-parent family', describes the Aimhigher pupils as 'a bit rough'. This reveals that she perceives these pupils as 'different' in terms of class. Carla's awareness of this may be rooted in her own sense of difference. This awareness, like that of the ambassadors on the extended degree programme, may be sharpened by being positioned herself as an outsider at Royal.

This resonates with Remi's account of the responses of 'young black boys'. Like Carla, she appears to be acutely aware of her and the pupils' 'difference' from the norm of white middle-class student at Royal. Like the women in Skeggs's study, it may be that Carla and Remi's awareness of their own positioning has contributed to their 'subjective construction' and to their navigation 'through classificatory systems' to 'evaluate themselves accordingly' (Skeggs, 1997: 4). As Reay *et al.* (2005: 99) found, 'class, linked to ethnicity, is ... central to fitting in and feeling comfortable' in HE. In this self-evaluation, though, there is a corresponding evaluation of the pupils with whom they are working and who are categorized stereotypically as 'lacking' the 'right' background (Burke, 2006; Thomas, 2001). This is in line with Remi's earlier comment about pupils, that 'they've never seen a young, black, professional male'. These accounts by the ambassadors emphasize pupils' learner identities

Social relationships and identities

– their G&T status – as being especially important. This may again relate to their own positioning at Royal, where their developing identities as successful learners, despite their difference in terms of race and class, provide them with a way of belonging at this elite institution (Reay *et al.*, 2009).

Ambassadors' dis/identification with these pupils may well be understood by them as 'a process of finding out what they cannot have' (Reay *et al.*, 2005: 85), and pupils may conclude that they do not belong to Royal. In contrast to the Aimhigher pupils, the G&T pupils on the summer school at Royal were accepted as appropriate by the student ambassadors. However, these pupils appeared to be from middle-class backgrounds with parents, uncles and brothers who were dentists, engineers, computer scientists, doctors and nurses. With their expectations of As and A*s at GCSE, unlike the Aimhigher pupils, they were a 'good fit' in academic and classed terms for Royal.

Friends and relations

In learning contexts with more 'informal attributes', male and female pupils frequently described ambassadors as being like friends and family members. Comparing ambassadors to family members was a particular feature of black pupils' descriptions, which may partly reflect how extended family networks are drawn on among Black African communities, described by Reynolds (2004) for Caribbean communities.

The pupils viewed their relationships with ambassadors as primarily social. The freedom to 'talk' was intricately linked to this. During conversations I asked pupils who ambassadors were most like; comparing ambassadors to family members was a frequent response:

> Sarah: We don't know them long enough to say they're like an older brother or sister, but maybe like a distant cousin.
>
> Dina: Yeah, like an older cousin that you can just talk to, yeah …
>
> Sarah: Not distant like far, far away but like – not like a cousin that you really talk to, but like a cousin that's like, yeah, at a family reunion or something.
>
> Joe: Just like a friendship.
>
> Bankside summer school

During the medical day, pupils again identified ambassadors as being like members of their family. One of the pupils in the morning group explained that they were 'like your family because they listen to what you have to say'.

133

Being able to 'talk to' ambassadors was repeatedly referenced by pupils and was clearly important to them. In busy school contexts, where the emphasis is always on providing information and checking knowledge, there is little opportunity for such talking and 'listening'. There has been a consensus in psychological studies that girls and boys 'do friendship differently' (O'Connor, 1992 in Hey, 1997: 10), and that the emphasis in feminine friendship is on 'its essentially private and intimate nature' (Johnson and Aries, 1983a and 1983b; O'Connor, 1992 in Hey, 1997: 10). These patterns, which were observed in this study as well, may reflect a reality in the hegemonic gendered expectations young people play out in their relationships with each other (Hey *et al.*, 2001). It is worth noting, though, that Davis (2001: 149), in his analysis of African American boys, observes that, despite appearances, the male pupils in his study crave a closer, more 'personal connection' with teachers, one in which they are listened to. Both male and female pupils discussed wanting to be 'listened to' during conversations; this 'listening' and the 'talk' pupils and ambassadors shared were repeatedly connected to ambassadors being seen as like 'friends' and 'family'.

Two pupils at the medical day described the difference in how being listened to by ambassadors and not being listened to by teachers makes them feel:

> Jenny: [They're] like a sister or a brother or someone, 'cause you can tell them what you're thinking and … whereas if you go to a teacher you can't – you feel more … because they know stuff.
>
> John: They are your teacher and you're here every day and they're just like – they make you feel really bad.
>
> Medical day

The pupils all shared the view that ambassadors were people who you could 'hang out with every day':

> John: But teachers, they don't listen to your point of view, but these people, they listen to everything … . Like a friend you would like to get to know.
>
> Abi: Like who you hang out with every day.
>
> Medical day

Discourses of friendships and relatives – of 'hanging out' – all position ambassadors clearly as having strong social relationships with pupils.

Social relationships and identities

During the G&T summer school at Royal, school pupils' response to ambassadors was unqualified enthusiasm – 'they're really nice', 'we really love them'. Pupils similarly described the ambassadors as like 'friends' and 'older brothers and sisters':

Lola: [Ambassadors are like] most people I know.

Clara: Friends.

Martin: Like older brothers and sisters ... like university students.

Imogen: Average teenagers.

Kate: ... they are like an older sister – a sibling.

G&T summer school

One pupil said ambassadors are 'like' her 'cousins'. This was because the ambassadors were understood by pupils as talking to them as equals, as being 'real about everything' and 'not trying to kind of protect us or baby us'.

One group of ambassadors discussed their social function as ambassadors to be to 'inspire children to chat to each other, make them bond' and to 'keep' pupils 'talking':

Munira: It's like conversations over lunch or like breaks, if it just bridges the gap or whatever, keep them talking, keep them in-tune ...

Alicia: I think we just chat; get them talking, really.

Marvin: And that's a nice way to start it off, 'cause once you get them start talking ...

Alicia: They don't stop.

Marvin: Yeah – once you open the floodgates ...

G&T summer school

The ambassadors discursively construct their work as facilitating the relationships within the group that they are leading. While these relationships are viewed as natural by school pupils, the ambassadors themselves are, to an extent, consciously facilitating relationships and encouraging social interaction.

Ambassadors also described their relationships with school pupils as like family:

> Munira: Older brother or sister, really. It's kind of informal but you still have an authority; you're not really like a teacher kind of thing.
>
> G&T summer school

Discursive constructions of ambassadors as being like friends and family were dominant during the two summer schools where ambassadors worked alongside school pupils for several days, but these discourses were notably absent among pupils during the engineering camp and maths workshops. While ambassadors and pupils similarly worked together for several days during the engineering camp and were again matched in terms of their backgrounds, the 'formal attributes' of both these contexts and the positioning of ambassadors as authority figures supervising and teaching pupils precluded the development of these social relationships (Gartland, 2014).

Performing identities
Small talk

Providing a social space where pupils and ambassadors can hang out and talk to each other was clearly important to the development of friendships and sibling/family-style relationships. It was within the small talk in this space that ambassadors and pupils had the opportunity to convey their cultural connections. Particular aspects of these conversations facilitated this process of identification.

Pupils repeatedly suggested that proximity in age enabled the ambassadors and pupils to relate to each other. Their accounts suggest that the types of conversation they had with ambassadors were different to those they have with other adults. One pupil explained that humour is important to this as 'older people wouldn't … find the humour in some of the things that we might find funny':

> Michael: And if you want to talk to them about a problem or you need advice or anything, because they've been there and they're doing that now only on a larger scale, so it's not like with the parents or older adults.
>
> Bankside summer school

Michael stressed the importance of the ambassadors' proximity in age to himself and their positioning as a fellow student – 'they've been there and they're doing that now only on a larger scale' – as enabling them to provide advice and information that he considered worth listening to. This positioning of the ambassadors was juxtaposed against that of 'teachers' and 'parents

Social relationships and identities

and older adults' who were viewed as less able to help. The significance of the positioning of ambassadors in these learning contexts is clear, since, despite pupils' perceptions, two of the ambassadors working with them were actually 30 years old, probably older than some of their teachers.

There were differences between the ways in which male and female pupils related to ambassadors, with girls finding it easier to bond quickly with ambassadors than their male counterparts. This was found during a discussion among pupils at the Bankside summer school about which sessions enable pupils to 'bond' with ambassadors. There was a clear gender divide in opinion. Dina argued that the afternoon sessions provide a better opportunity, as the pupils and ambassadors share a common interest. Sarah agrees with her, but this was refuted by Michael and Joe, who saw the longer-term relationships developed with the engineering ambassadors during the morning sessions as better for facilitating conversations:

> Dina: Yeah, when we do our chosen activity you bond more with people because they are doing that activity that you have chosen to do, so you have more in common 'cause you know that you both have …
>
> Sarah: Oh yeah, that makes so much sense.
>
> Clare: So in the afternoon session – go on, say that again.
>
> Dina: In the afternoon you bond more with them, as you have both chosen to do that activity – they are, like, leading it, so you have a bond – a common …
>
> Sarah: You know that they enjoy the same thing.
>
> Dina: Yeah, so you can start a conversation, whereas on this one you don't know particularly if they have chosen to do it or …
>
> Clare: So you have more – you know, you talked about the different conversations you've had – do you have more conversations just about life and stuff in the afternoon?
>
> Dina: Yeah, where in the morning …
>
> Michael: What, with the student ambassadors? I think we have more conversations with the student ambassadors in the mornings. 'Cause then they are like the engineering ambassadors that we saw every day …
>
> Bankside summer school

Pupils found it easier to develop relationships with ambassadors if they had a common interest, and longer-term relationships facilitated trusting relationships. In the discussion, though, the male and female members of the group prioritized these differently. This reflects my observations that male pupils were less likely to initiate conversations with ambassadors than their female counterparts and took longer to build relationships. By contrast, female pupils started conversations more often and developed relationships with ambassadors relatively quickly. This may reflect girls' orientation to 'essentially private and intimate' relationships with each other (Hey, 1997: 10).

During the G&T summer school, the desire to have fun and socialize was pointed to as a shared aspect of the ambassadors' and pupils' identities:

Lola: They're no different to us.

Kate: They don't just, like, want to work all the time, like, they want to have a life as well.

G&T summer school

Among this group of pupils, socializing appeared to be linked to drinking, which was referenced implicitly and explicitly during my conversation with them. Martin describes conversations with ambassadors as 'normal chat' and describes this as 'like you would chat to a friend, obviously, about the bars and stuff'.

Ambassadors at this event identified various topics of conversation including clothes, film, TV, Facebook, Michael Jackson (who had just died), sport and food:

Marvin: Football's about it, to be honest.

Than: I talked to somebody about cricket …

Carla: I talk about food a lot …

Munira: That's how we bond with them; we talk about TV shows and music to get them talking to us … heroes – I don't watch much – they just think that I've heard of it; I don't watch TV or listen to music, so I'm pretty much screwed, so I am like, yes – Transformers yes – Batman maybe – I just kind of wing it really …

Than: You talked about films at lunchtime.

Munira: I was guessing everything …

Social relationships and identities

Alicia: I do the clothes thing though – or they tend to do the clothes thing ...

Carla: I've been asked about holidays, plans for the summer, future, jobs ...

Marvin: Just their hobbies and interests as well, whether they play on football teams and stuff like that ...

Munira: I think we just chat; get them talking, really.

Marvin: Don't really think about it, do you – just talking.

G&T summer school

Though most appeared to find this fairly 'natural', ambassadors were clearly consciously working to find topics of interest to engage pupils, and these topics referenced normative gendered student and pupil identities. Munira's perception that her lack of knowledge of TV and music meant that she is 'pretty much screwed' reveals the centrality of these topics for conversations with pupils. Female ambassadors identify food and clothes as topics, while male ambassadors identify football and cricket. These topics are clearly gendered and, to an extent, ethnically and culturally based, with this group of white pupils referring to practices of bars and drinking that are not social practices shared so widely with black south London teenagers.

During the Bankside summer school, pupils similarly described general conversations about 'things' with ambassadors. These were lighthearted conversations that pupils also have with peers that again reflect their gendered and cultural locations. It may be that the ambassadors' ability to relate to pupils about such 'things' is more important in pupils' perception of ambassadors as being 'young' than their actual age. The subjects identified particularly related to music and girls' appearance:

Sarah: You can just ask them like, things like – 'how was your day' – or 'my god, did you see that girl's hair?'

Clare: Anything else you talk about?

Stacy: Music.

Bankside summer school

A shared knowledge of music featured a number of times in the exchanges between ambassadors and pupils during the Bankside summer school. The importance of hip-hop culture to black male identity is well documented (Kitwana, 2002; Reese, 2004; Gosa and Young, 2006). Hip hop was a frequent topic of conversation, particularly during the Bankside summer school. My observations indicate that during the summer school, ambassadors were perceived by school pupils to identify with and be knowledgeable about hip hop and that this contributed to pupils' identification with them.

During a conversation with pupils, I reminded one girl that she had talked about having conversations with ambassadors about hairstyles – 'my god, did you see that girl's hair?' She explained how topics differed between conversations with female and male ambassadors:

> Sarah: No – the female ambassadors ... but with the male ambassadors you just talk about stuff – like general conversation – like, just stuff – like, 'what football team do you support' – you argue with them, 'cause they probably say Manchester United and I say Arsenal and then I argue.
>
> Bankside summer school

These gendered conversations were repeated over the summer school. Two female ambassadors and two pupils were talking about how easy it was to spend credits on their mobile phones. One ambassador said that she had spent all her wages 'on a dress and some shoes'; she said to the female pupil 'that's what students spend their money on'. On another occasion a pupil initiated a conversation with an ambassador about her looks saying that Freya looks young 'the way you do your hair' and that she was 'really surprised' to find out she was older, saying other ambassadors and adults in the room 'dress older'.

This patterning of conversations was repeated during the medical day with Simi, a female ambassador, commenting that it was pupils who initiated conversations about her appearance:

> Adi: Yeah, definitely – with a boy you can talk about football – the latest transfers, what team's the best and stuff like that and I think if they saw me as a teacher, as a member of staff like Doctor R, I don't think they would start off a conversation like that – I think definitely ...
>
> Simi: Yeah, like a girl asked me the other day, 'where did you get your dress from?' Hello science, welcome to my clothes ...
>
> Medical day

Female pupils closely observed ambassadors' appearance and that of other girls, resonating with other accounts of female friendships, especially in psycho-social research (Hey, 1997; Renold and Ringrose, 2008). Girls

Social relationships and identities

'draw boundaries' about how it is acceptable to look and are 'placing others outside those boundaries' (Hey, 1997: 33). As Hey explains, it is through such processes that 'we establish our identities'. The instances outlined here illustrate moments where pupils are in the process of aligning their own gendered identities with those of ambassadors and even othering other girls in their peer group in the process, illustrating the centrality of gender to identification processes.

Physical interaction
Appearances

Butler (1990: 151) coined the term 'heterosexual matrix' to explain the 'grid of cultural intelligibilities through which bodies, genders and desires are naturalised'. This is useful for looking at enactments of (gendered) identities in the pupil–ambassador relationship.

When asked about similarities or differences, pupils on the G&T summer school focused exclusively on ambassadors' similarities to themselves, including their being 'friendly', 'relaxed and casual', 'being fun' and being young – 'they're only about 19'. As discussed, the fact that ambassadors and pupils are all learning seems part of this youthful identity – 'there's no-one that knows everything yet'.

Ambassadors as 'fun' was a recurring theme in pupils' accounts. The youthful identities of ambassadors and their ability to relate to pupils as 'equals' was significant to their being seen as fun. This in turn contributed to them being viewed as fun to socialize with, to go 'out for a party' with:

> Lola: Oh, they're really nice.
>
> Martin: We really love them and get on with them.
>
> Imogen: 'Cause they treat us like equals.
>
> Lola: And not like some little students.
>
> Martin: We like them so much that yesterday we went out for a party (*inaudible*), but they are really nice people and obviously they are not too much different in age.
>
> G&T summer school

The positioning of ambassadors in these social locations was intimately connected to the type of event they were working at and whether they had the opportunity to develop relationships, even friendships, in these learning contexts. Pupils' interest and, in some instances, identification with ambassadors was closely connected to how ambassadors presented themselves, both through their talk and through their appearance. That pupils perceived ambassadors to share their sense of 'style' was clearly important to pupils in the process of their gendered identification. One pupil's explanation that the ambassadors are 'like us, but older' was closely linked to this issue of style:

> Clara: They wear normal stuff – they are normal people. They just wear the same stuff as us really. They are like us, but a bit older.
>
> Medical day

Observations and conversations about dress were largely held by female pupils and ambassadors. This 'surveillance' (Foucault, 1977: 216) among girls has been well noted (Hey, 1997). Davis (2001: 147) explains that among black boys there are also clear rules about 'what's accepted masculine presentational behaviour'. Both male and female pupils were performing gendered identities that are acceptable within the 'heterosexual matrix' (Butler, 1990). That medical students and engineering students are 'surveilled' and seen as acceptable is a vital part of the identification process and one that appeared to take place during social interactions at these events.

These identifications may also 'rupture' (Renold and Ringrose, 2008) or at least destabilize some younger pupils' ideas about what are possible future identities for themselves. Black male ambassadors who represent 'standards for hip-hop culture' and conform to 'accepted masculine presentational behaviour' (Davis, 2001: 141–7), but are high academic achievers and not part of an 'oppositional culture' (Fordham, 1996) provide pupils from low-achieving schools with 'viable ways of being' (Davies, 2006: 430). However, it must not be forgotten that these pupils already had established strong learner identities.

Another 'rupture' to the 'heterosexual matrix' was the performed identities of the female engineering ambassadors. In her article about 'engineering identities', Walker (2001: 81) describes how female engineering students 'construct themselves as in some way "different" from other girls'. The girls in her study emulated the behaviour of their male counterparts – a type of performance described by as 'licensed mimicry' McRobbie (2006: 10). However, the female engineering students in this study seemed at ease with the dual positioning as appearing to conform to hegemonic gendered identities and being engineering students. An image that remains with me from the Bankside summer school is of Qadira, a black female engineering student, during an end of course party, dressed in a sparkly white fairy outfit with wings, laughing and joking with pupils as she served them pizza. This performed identity interestingly locates Qadira within normalized feminine

Social relationships and identities

'girly' identities, and yet this performance within the context of such a male-dominated subject area provides a challenge to the ascendancy of dominant masculine identities – Qadira was certainly not taking up male student identities in the way that Walker (2001) describes in her study.

Physical contact and issues of child protection

There was a clear contrast in the amount of physical contact between ambassadors and school pupils during the two summer schools. Physical contact between teachers and pupils is both regulated and subject to 'surveillance' by teachers, parents and pupils (Jones, 2004). However, physical contact is an expected expression of friendship within peer groups, though there are clear gendered boundaries relating to what is permissible within the 'heterosexual matrix' (Butler, 1990: 151). Ambassadors' positioning in some contexts as like friends, siblings and cousins led to pupils responding to them in physical ways as they would to friends and relations.

During the summer school at Bankside, physical interaction between pupils and ambassadors increased as relationships developed over the week. Pupils and ambassadors often exchanged high fives. There were times when pupils hugged ambassadors. I also noted an exchange between a male pupil and a male ambassador. The pupil was cross and frustrated, as the batteries in his robot had run out and he could not continue to program it. Dan, the ambassador, checked that this was true; the pupil watched, frustrated, saying 'no, it's out of batteries' and Dan responded by laughing and massaging the pupil's shoulders saying 'it's alright, it's alright'. This diffused the situation and the pupil looked up at Dan and smiled.

I noted physical exchanges only between pupils and ambassadors of the same gender. Such physical interaction between male ambassadors and female pupils or vice versa would, within the 'heterosexual matrix', constitute a sexual advance. This appeared to be an understanding that ambassadors and pupils shared. The lack of physical interaction between the opposite sexes is indicative of how conscious pupils and ambassadors were of their gendered and heterosexual positioning. If, however, a young male teacher had massaged a male pupil's shoulders during an interaction, this is likely to have been construed as 'suspicious' (Jones, 2004). That these gendered physical interactions were welcomed and often initiated by pupils is indicative of how far from 'teacher' and how much closer to 'friend' they perceive their relationships with ambassadors to be.

However, these physical interactions were not so evident at the G&T summer school at Royal – indeed I observed no physical interaction at all. A discourse drawn on during the focus group with Royal ambassadors during the

G&T summer school was related to child protection. Issues relating to child protection had been covered quite extensively during ambassador training, as the WP coordinator leading the session had worked at a Young Offenders unit and drew on her knowledge of a range of child abuse cases. She stressed the importance of not giving out any contact details, of avoiding physical contact and never promising to keep confidences. The ambassadors discussed the practicalities of abiding by these 'rules' during the focus group. One area of concern was 'being alone' with pupils, something the ambassadors had been warned to avoid, but that they discussed as problematic. Than described an elaborate subterfuge he initiated to avoid such a circumstance:

> Than: I feel really weird about the whole – like you're not allowed to be with them alone sometimes and it's so, like, annoying … I was telling you today, wasn't I, that I turned up early, like nine, and Dan was here and I was like, 'oh, you're here early', and then I went to check the room and I was like – 'it's empty, you can come in', and then I thought, oh, one minute, it's just me and him – this isn't allowed, so – and then I was like, 'oh I just need to go – I just need to go' – and then he said, 'oh, that's alright, I'll come with you', so then I went, 'okay', so then we walked upstairs and then I just said, 'oh, I need to go to the library', 'cause I knew he couldn't swipe in …
>
> Munira: This is like mission impossible.
>
> Than: So then I just went into the library and checked my e-mails for about ten minutes and came back and he was waiting for me and I felt really, really bad … 'cause he was just sitting there for ten minutes, but I didn't know what else – I couldn't think of what else to do and by the time we came down there were other people here so it was fine. I haven't had anything like that before in my life.
>
> G&T summer school

Than's consciousness of abiding by the rules left him feeling 'really bad' as he left a pupil sitting on his own for ten minutes. Indeed, such practices may actually leave pupils feeling rejected.

Another problem related to child protection were the rules regarding contact with pupils following events:

> Alicia: No Facebook.
>
> Carla: No contact details.

Social relationships and identities

Munira: No e-mail, basically no contact after …

Clare: And how does that work – I mean do people …

Alicia: It's hard.

Munira: You just have to …

Than: They say no e-mails and stuff and I think Facebook – there was a rule.

Alicia: No Facebook – ignore or reject.

Than: They say that – there's no like, 'you definitely can't do it' but it just saves your back.

Marvin: Better safe than sorry.

Munira: … you can't actually give the details to that person because you can't get in touch with them and at other times I've said, 'oh can you pass this on' and they haven't – like, to the person that you meant to – and that's annoying, 'cause I look like I've not actually done what I've said I am going to do, but it's not that, and I think it's quite weird, because they don't know about child protection.

G&T summer school

The ambassadors shared Marvin's view that it is 'better safe than sorry', but this ban on contact evidently posed problems. Munira explained specifically how she had been prevented from passing on details about events to pupils, because she had been unable to contact them herself and the contact with pupils via the university had proved unreliable. She also suggested that pupils 'don't know about child protection' and may not understand the ambassadors' behaviour. Ignoring or rejecting pupils' invitations to join Facebook, for example, could certainly be viewed as a rejection of the friendship.

Ambassadors' discussion of the rules about not touching pupils shows that they were also seen as problematic. Munira described how she 'pats' pupils to encourage them in their work and how pupils at times 'come over and hug you'. Another ambassador described how a pupil was 'touching' another ambassador's leg. Than even explained how he is reluctant to 'shake hands' with pupils:

145

> Munira: Apparently, there's no physical touching.
>
> ...
>
> It's difficult, 'cause sometimes they come over and hug you and you're like – oh, what are you meant to do?
>
> Carla: Natalie [a pupil] was a bit inappropriate – she was, like, touching Bim's [an ambassador] leg. On the first day she was going, 'oh you're so cold', and she was like – 'ooh, don't touch me'.
>
> Alicia: Guys don't get much touching.
>
> Than: No, but, like, a few times someone's shaken my hand when we go and I'm not going to say, 'I'm sorry, I can't shake your hand'.
>
> Girls: You can shake hands!
>
> Munira: I think patting is alright as well; it's just not caressing or, like, hugging ...
>
> Carla: So where are the boundaries really?
>
> Alicia: Yeah, the boundaries; it's fair enough to say no physical contact, but what does that actually mean?
>
> G&T summer school

These rules are problematic for ambassadors, as they conflict with ambassadors' positioning as being 'like a friend' or an 'older sibling'. In such relationships, light physical contact, keeping in touch and chatting alone in a room together are all natural and expected. Ambassadors' avoidance of these expected aspects of developing relationships may undermine relationships that they have worked hard to build.

Challenging identities?

The work of ambassadors with pupils both challenges existing patterns of participation in HE and also reinforces them. In some contexts, pupils appear to identify with ambassadors, although this process of identification is complex. It is simplistic to assume that ambassadors are necessarily aspirational role models for pupils just because they are male, female, young, students or black. In this study, ambassadors who are viewed positively by pupils in one learning context are at times subject to criticism and even viewed with hostility in another. However, the influence of intersecting (gendered) identities of ambassadors and pupils, combined with acting out idealized engaged and motivated student identities in a social space created by collaborative learning contexts, was at times powerful. School pupils repeatedly joined in this performance during activities and aligned themselves socially with student ambassadors. This performance may be part of a process of trying out 'viable ways of being' (Davies, 2006: 430) or of the 'naming and making' of future university students (Youdell, 2006). However, there are also instances, particularly in learning contexts with more formal attributes, where differences in pupils' and ambassadors' intersecting identities were highlighted. In these contexts, pupils' interactions with ambassadors reinforced pupils' sense of difference. These interactions can entrench pupils' identities in opposition to the student and subject identities of ambassadors.

Also, as Butler (1997a) suggests, there is a 'restriction in production' and this is a central issue for the WP aims of these projects. The participation of black and working-class pupils from south London schools in activities at Bankside may serve to reinforce existing patterns of HE participation, as they find Bankside a comfortable environment where they are surrounded by people like themselves. This process may actually serve to preserve the status quo rather than challenge it; likewise, the attendance of white middle-class pupils at the summer school at Royal. Constraints on the friendships are also imposed by issues of child protection, and these constraints not only impact on the friendships that develop during activities, but also on the opportunities for interaction between ambassadors and pupils after events and on the potential ambassadors have in supporting pupils who do progress to the university.

There are, however, instances of 'rupture' and disruption'. The juxtaposition of sparkly female and engineering and science identities is one such instance, the bringing together of hip-hop culture on the one hand and, on the other hand, science and engineering identities among black male student ambassadors is another. These ambassadors serve to provide a challenge to assumptions about science and engineering identities. Pupils are seen to take up, if only for a short time, these different ways of being. These interactions have the potential to 'interrupt dominant identity patterns of (dis)identification' with STEM (Archer et al., 2010: 21).

Notes
[1] Reflecting trends identified in the study of petrochemical engineering in the UK among Black African women outlined in chapter 3.

Chapter 8

Assumptions, practices and potential

When I started this study, I had been evaluating WP initiatives for several years and had observed the strategy of HEIs to employ student ambassadors in 'aspiration raising' work. While I had observed that ambassadors are often well liked by pupils, I was interested in unravelling what was actually happening in these relationships and how this activity connected to wider policy imperatives and to redressing inequalities in HE access. At the onset, I had a number of questions about the work of ambassadors. I was interested in exploring the nature of the relationships they formed with school pupils, the significance to these relationships of different aspects of the pupils' and ambassadors' identities, including learner and subject identity, gender, ethnicity and class, and the learning that takes place within these relationships. I was particularly keen to interrogate assumptions that ambassadors automatically become role models for school pupils. Most importantly, my aim was to critically scrutinize assumptions that ambassadors contribute to widening participation and progression generally and in STEM subject areas in particular, specifically in engineering and medicine.

I was keen that my work should be supportive and of practical use to WP practitioners and should provide useful insights into how ambassadors can effectively support the equality and social justice agenda that genuinely motivates practitioners in this field. I also thought it was important that, amid the frenetic activity and drive to meet targets, my research should provide a space in which pupils' voices could be heard, or more specifically the voices allowed by the dominant discourses and different pedagogies and practices employed in these outreach activities. I felt that such an exploration would also contribute to wider ongoing sociological and policy debates about progression in education and young people's decision-making.

The methodological approaches used in this study, my engagement with a range of theoretical frameworks and analytical tools have enabled me to consider schemes through both the wide-angle lens of policy and globalized discourses, and through a magnifying glass, to observe the minutiae of interaction between participants. I believe these joint perspectives have provided insights that are useful to practitioners, while also critically scrutinizing these WP initiatives and posing questions and issues for academics and policy makers.

Findings were organized into three chapters reflecting dominant discursive constructions and discourses in the accounts of participants. The first focused on dominant neo-liberal marketing and related discourses, the second on learning, pedagogy and related identities, and the third on intersecting aspects of pupils' and ambassadors' identities, how these contributed to dis/identifications and how interactions can be performative, forging identities in relation to HE and subject areas

Meanings of marketing

Marketing discourses circulating within both institutions were powerful and dominant, reflecting a neo-liberal regime of truth which positions pupils as consumers within the marketplace of HE. Student ambassadors were quite clearly enacting this discourse, marketing university generally and their own institutions in particular. At Royal, the ambassadors' work, organized by the WP unit and funded by Royal, targeted G&T pupils. If the summer school is taken as an example, many of these pupils were from established middle-class backgrounds. Bankside ambassadors were working with a range of local schools with large numbers of ethnic-minority pupils. These local working- and lower-middle-class and minority-ethnic groups are already over-represented at Bankside and at other similar new or WP universities. A generalized marketing focus across subject areas therefore seems more likely to reinforce existing patterns of participation at such universities than to challenge them. This fits into a wider neo-liberal discourse of the knowledge economy, where political strategy has been focused on upskilling the UK workforce to increase competitiveness within global markets. The central ambition of this strategy has never been to change patterns of participation within the HE system, but simply to increase it.

During the years of New Labour administrations, discourses of the knowledge economy were in constant circulation. There was a media and political focus on the graduate premium, arguing that anyone attaining a degree is likely to be significantly better paid than their less-educated peers. However, Morley (2001), Lauder *et al.* (2008) and David *et al.* (2010) suggest that policy assumptions building on the idea that there would be graduate jobs for all of these young people to progress into, would be unlikely to hold true. WP strategies that simply encourage more young people to progress to university generally may be selling young people a future career that is, in reality, unattainable. Current, well-documented difficulties faced by some graduates in attaining graduate jobs certainly support this view.

At Bankside, during the time of the study, discourses of employability (Morley, 2001; Brown and Hesketh, 2004; Moreau and Leathwood, 2006; Hey and Leathwood, 2009) were dominant. The ambassadors' 'role' was becoming increasingly professionalized. This may have benefits for school pupils in terms of continuity of ambassador engagement and increased reliability and commitment. However, I would argue that there is a danger that the dominant discourses of marketing combined with the emphasis on employability and professionalism could result in positioning ambassadors as marketing professionals. School pupils were aware that ambassadors were promoting university and subject areas. If this focus on marketing university and courses is too overt, it is likely to undermine the trusting relationships (Ball and Vincent, 1998; Ball *et al.*, 2000; Archer *et al.*, 2003) that ambassadors and pupils can develop during their work together (Gartland and Paczuska, 2007; Gartland, 2012/13).

What, then, of the work of student ambassadors within particular subject areas? The ambassadors in this study were consciously promoting engineering or medicine. The medical student ambassadors promoting the MAS at Royal appeared to be effective in this. With many coming from the same schools and backgrounds as the pupils in the south east London schools targeted, their promotion of Royal and medicine resonated with school pupils. This programme was clearly effectively enabling groups of young people traditionally excluded from this subject area to study medicine. However, there were aspects of this promotional activity that concerned me. The pupils on the G&T summer school appeared to come from predominantly white and middle-class backgrounds. So, while successful, this promotional activity was not likely to impact on classed or ethnic patterns of participation in medicine. The ambassadors promoting the MAS themselves were evidently pleased to be at Royal studying medicine. However, this had a psychological cost (Reay *et al.*, 2005; Crozier *et al.*, 2010). The accounts of this group of ethnically diverse pupils from low-achieving comprehensive schools in south east London suggested that they were marked out as different to other students at Royal. These differences contributed to discourses of deficit and charity surrounding their work (Thomas, 2001; Burke, 2002; Archer and Yamashita, 2003; Yorke and Thomas, 2003; Watts and Bridges, 2004; Bridges, 2005) that were not present within the accounts of ambassadors at Bankside. This reflects research that has found pupils progressing to elite institutions from working-class and ethnic-minority backgrounds have to undergo a 'fracturing of identity' (Reay, 2001; Crozier *et al.*, 2010). The findings of Crozier *et al.* seem to resonate particularly with the student ambassadors' accounts in this study. They identify the differences in the experiences of working-class students at different institutions. For the students at the elite university in their study, their social positioning was uncomfortable; academic work and 'being a student became their main source of identity' (Crozier et al., 2010: 70).

Another concern has been that ambassadors were enacting another neo-liberal discourse, that of individualism, which obscures the reality of the structural obstacles facing pupils. Their accounts focused on the power of the individual (Beck, 1992; Ball et al., 2000) to surmount the difficulties presented by having attended a low-achieving school. These ambassadors had themselves progressed to an elite institution to study medicine; they were engaged and motivated students and therefore motivated by the task of encouraging other young people from similar backgrounds to follow the same path. However, the reality of available routes into HE and medicine in particular is not so clear for these school pupils.

The extended MAS offers a small number of places to a small number of pupils who achieve Bs rather than As in their A levels. These places have now been opened up to pupils from a far wider catchment area. The opportunities to achieve a place are extremely limited. As Delgado (1991) points out, this promotion of a career that is, in reality, inaccessible to most pupils from economically disadvantaged backgrounds, is problematic. While the work of ambassadors on this scheme did enable a small group of pupils from deprived neighborhoods to study medicine, it is not going to change patterns of participation on medical degrees at Royal as a whole. Indeed, this discourse of individualization, of opportunities for the 'deserving poor', is not new and has long traditions that are reflected in the history of the charities supporting the MAS programme.

The focus of the work of the engineering ambassadors at Bankside was more ambitious than the MAS. Chapter 3 outlined how engineering courses across the UK attract far fewer applicants than medical degree courses, with some struggling to recruit sufficient appropriately qualified applicants. Various government and industry reports have stressed the importance of STEM subjects to the UK economy as a whole (Leitch, 2006; Lambert, 2003; Sainsbury, 2007; DIUS, 2008; Perkins, 2013). The CBI (2010) report states that 45 per cent of employers were having difficulty recruiting STEM-skilled staff at the time of the report, with almost six in ten, or 59 per cent of firms expecting difficulties in the following three years. The ambition of the AEP was in part to promote engineering as a possible career to WP cohorts of pupils and to female pupils. As well as routes into HE, the AEP also promoted other routes into engineering via work-based learning and apprenticeships. There are well-paid jobs available, for example, within the service sectors, which are anxious to recruit people who are representative of the communities they serve. There is also the potential to work as graduate engineers for international companies. Engineering could provide young women, minority-ethnic and working-class pupils with well-paid careers that would not necessarily require them to leave their communities. There are also jobs and training opportunities at different points in young people's learning careers, not exclusively for graduates. The gender imbalance in engineering has been an identified problem for many years, but remains entrenched. The promotion of engineering, then, in comparison with broader WP aims to increase HE participation generally, seemed to be better linked to actual opportunity. However, it is worth observing that this conception of opportunity functions within the confines of the discourse of the knowledge economy and neo-liberal interpretations of HE as existing to provide suitably qualified graduates for the benefit of the economy.

What, then, do student ambassadors contribute? Working with ambassadors in some contexts contributed to endorsing and consolidating interests in engineering and STEM subjects. But, as Archer *et al.* (2010) identify, pupils' orientations to subjects are often established early on in their learning careers and before Years 10 and 11, the age groups targeted by several of the interventions I observed. Engagement with ambassadors in some contexts was successful in promoting engineering messages and raising pupils' awareness of engineering careers. Where ambassadors and pupils were working in contexts with more attributes of informality (Colley *et al.*, 2003: 30–1), alongside each other and engaged in practical engineering tasks, ambassadors' messages about job opportunities were heard. This is significant, as pupils' hierarchical perceptions of courses impacts on their HE choices (Brooks, 2003b). Ambassadors in these contexts also successfully conveyed the message that engineering requires creativity, though it was unclear how well the shorter activities actually provided pupils with an understanding of the technical skills and knowledge involved.

Learning practices and identities

At the outset I was keen to focus on the learning that takes place between ambassadors and school pupils. Among WP practitioners, there seems to have been no clear or shared vision of what ambassadors are there to achieve. The audit culture (Colley, 2005) in education generally and the focus within widening participation on proof of impact has led to a preoccupation with numbers and progression into HE. This focus has detracted from useful exploratory research into the nature of the learning that takes place. I was interested in exploring what it is that pupils are learning from their interactions with ambassadors and to draw on learning theory in this exploration. I traced discourses relating to teaching and learning in the various contexts that I observed. This research reveals that discourses were notably different in different contexts.

Drawing on theories relating to formal and informal learning and experiential learning has provided tools with which to conceptualize the difference between the learning environments and the particular roles that ambassadors were allocated. The focus within schools on the league tables and credentialism (Gillborn and Youdell, 2000; Williams *et al.*, 2010) has led to schools requesting curriculum support from ambassadors. However, contexts where ambassadors were placed in more formal learning environments and positioned as teachers were problematic. They were not trained as teachers or in teaching the curriculum and were not able to replicate the work of teachers. Also, when ambassadors were positioned in authority over pupils, it was easy to see how they could develop the authoritarian personalities discussed by Chilosi (2008) or the polemic view of 'bad' pupils versus themselves as 'good' students found in Taylor's research (2008). In these contexts it was very difficult for ambassadors to develop relationships with pupils in which pupils view them as trusted hot sources of information. In learning contexts with more formal attributes, the ambassadors were certainly not seen as role models by the pupils they worked with. Indeed, the research indicated that ambassadors' subject expertise could act as a barrier, with pupils dis/identifying with ambassadors on the basis of their subject identities.

Contexts with more informal attributes, where ambassadors and school pupils worked collaboratively, however, allowed for relationships to develop between ambassadors and pupils. In a few contexts activities were carefully planned and provided pupils with insights into real-world applications for subjects. These activities drew on practices within HEIs of experiential problem- or project-based learning within medicine and engineering. In these contexts, pupils and ambassadors worked collaboratively and were both positioned as learners, though at different stages. In this situation, ambassadors were much more successful in developing relationships with pupils where they were viewed as trusted hot sources of information. However, it is important to further interrogate the quality of the information that ambassadors provide when positioned as sources of information, advice and guidance in learning contexts. Their own experience alone does not adequately equip them for this responsibility. There is also the danger that ambassadors' reliance on their own experience provides individualized success stories that are misleading or even alienating for pupils. The students on the MAS programme may successfully enthuse pupils to aspire to study medicine, but this emphasis on individual and relatively isolated success obscures the reality of the structural obstacles that pupils have to overcome and the limited options open to them (Ball *et al.*, 2000; Archer *et al.*, 2003; Reay *et al.*, 2005). There is also the danger that the promotional focus of ambassadors' interactions with pupils actually contributes to opacity (Ball *et al.*, 2000: 10) rather than clarity in terms of the information they provide about HE.

When considering the relationships that develop in these more informal learning contexts, I would suggest that it is useful to consider Hodkinson and Macleod's proposition of 'learning as becoming' (2007). Shared, practically oriented activities between ambassadors and pupils provide brief periods of time where student, subject and learner identities are 'constituted through action' (David *et al.*, 2006: 422). During these periods of shared and collaborative activity pupils learned both about STEM subjects and also about becoming students.

What is significant in my findings is that the learning environments within which pupils and ambassadors are placed and the positioning of ambassadors within these contexts are influential over the relationship that develops. Where contexts facilitated collaborative working, ambassadors and pupils were both positioned as learners. This positioning was effective in enabling ambassadors and pupils to develop a shared student identity that may even develop and constitute (Butler, 2004) pupils' own student and learner identities (Gartland, 2014). It is worth noting, though, that pupils who already have established identities as learners are more likely to identify with ambassadors in this way. Pupils' identities as high achievers are established early on in their learning careers (Brooks, 2003a; Crozier *et al.*, 2010; Reay *et al.*, 2009). These identities impact on HE aspirations (Brooks 2003a) and on their success within the HE system (Reay *et al.*, 2009; Crozier *et al.*, 2010). Ambassadors may help to support and encourage these identities, but if pupils do not have established learner identities, this support and encouragement is less likely to be effective. It is worth noting here that many of the school pupils in this study were from G&T backgrounds or were already positively orientated to HE. One of the ways in which learning is viewed as being situated in work-based learning is in terms of the 'socio-biographical features of the learner's life' (Evans *et al.*, 2006: 13). Positive learner identities are highly significant in the ambassador–pupil relationship and, as Taylor's (2008) findings indicate, placing ambassadors in learning contexts with more formal attributes (Colley *et al.*, 2003) with pupils who do not have positive learner identities can actually serve to reinforce their sense of difference.

In terms of developing STEM identities specifically, ambassador work has potential. Pupils appear to respond most when they are already positively oriented towards STEM subjects. In these instances, the shared identities that pupils develop through working with ambassadors reinforce and embed pupils' identification with STEM subjects.

Social relationships and identities

Generalized discourses about ambassadors being role models circulate widely within educational institutions, but there is no shared understanding of how this plays out in practice. The term 'role model' has become part of conventional public and public-policy discourse and is used ubiquitously, especially in relation to WP work. All those involved in the administration of ambassador work, as well as the ambassadors themselves in this study, drew on this discourse. One way in which this discourse was taken up and enacted by student ambassadors was in marketing their own institutions and courses. Another field of meaning drawn on was more specific and focused on how ambassadors were modelling behaviour, such as being engaged, focused and motivated, that it was hoped pupils would emulate. This positioning of ambassadors is useful and has the potential to encourage pupil identification with ambassadors.

In contexts with more informal attributes (Colley *et al.*, 2003), pupils viewed the relationships they had developed with ambassadors as predominantly socially oriented. This was illustrated by pupils' descriptions of ambassadors as being like friends, cousins and brothers and sisters. While both male and female pupils described ambassadors in this way, the comparison to family members was found particularly in the accounts of black pupils which may reflect more extended family networks (Reynolds, 2004).

My findings here illustrate that these socially oriented relationships are quite delicate and easily destroyed or undermined. Descriptions of ambassadors as being like friends, cousins or brothers and sisters were notably absent in contexts with more formal attributes, where ambassadors were positioned as didactic teacher or authority figures. Indeed, in these contexts there were instances where pupils were consciously dis/identifying with and othering ambassadors (Hey, 1997). In one instance, there was an almost exuberant defiance, in pupils' accounts, of the respectability (Skeggs, 1997) associated with one white middle-class female ambassador. Or, perhaps, as Davies (2006) found in her study of boys in primary school, these pupils may have been subverting power relations by taking up positions that provide alternative successful black female identities.

Discourses of deficit also affected these relationships. Ambassadors at Royal whose own raced and classed identities were different to the white middle-class norm at the institution appeared to be particularly conscious of pupils' backgrounds and were quick to identify them as lacking the necessary knowledge needed to access Royal. As Skeggs (1997) described, 'rough' working-class identities are compared here to the 'respectable' identities of the middle class. Central in ambassadors' accounts was that these pupils did not have appropriate learner identities (Reay *et al.*, 2009). These discourses of deficit were, as discussed, not found at Bankside, though pupils and ambassadors shared similar backgrounds to those found lacking at Royal.

As has been suggested in various studies of mentoring relationships (Klaw and Rhodes, 1995; Sanchez and Colon, 2005; Cavell *et al.*, 2002; Jackson *et al.*, 1996; Kalbfleish and Davies, 1991; Sanchez and Reyes, 1999; Chen *et al.*, 2003; Ensher and Murphy, 1997), the backgrounds and identities of pupils and ambassadors are important in terms of how they relate to each other. Pupils' responses to ambassadors revealed how their gender, ethnic, class and cultural backgrounds all intersected to inform how pupils related to them. The dis/identification observed was intricately linked to pupils' and ambassadors' intersecting (Crenshaw, 1989; Mirza, 2008; Morley, 2012) gendered, ethnic, classed identities, as well as their learning identities.

The data presents a complex picture of how aspects of pupils' and ambassadors' identities impact on their relationships. Like Hockings *et al.* (2010: 195), I have tried to present a multifaceted view of student diversity which extends beyond the structural relations or divisions of class, gender and ethnicity and have attempted to explore the interplay of different aspects of pupils' and ambassadors' identities (Liang and Grossman, 2007).

The widely accepted notion that pupils will aspire to be like ambassadors just because they are young, black and male or female is clearly simplistic. There are many aspects of ambassadors' identities that pupils are aware of and respond to. There were instances in this study where pupils identified closely with ambassadors. Ambassadors' positioning within contexts, as well as their gendered and other intersecting aspects of their own identity and those of pupils did seem important, though there was no clear formula to how this worked in practice. Similarity of interests and attitudes were clearly important in some instances (Ensher *et al.*, 2002; Ensher and Murphy, 1997; Grossman and Rhodes, 2002). Significantly, a shared conception of themselves as learners contributed to pupils' identification with ambassadors.

I drew on Butler's theories of performativity (1997a) to explore pupils' learning during different activities. Where ambassadors were placed in authority over pupils and in didactic roles, this encouraged dis/identification; if these learning experiences with ambassadors are seen as performative, they could even damage emerging HE and subject identities. In contrast, in practical learning contexts, ambassadors were performing 'ideal student' and pupils were positioned as fellow learners, and joined in with this performance. Pupils were more likely to identify with ambassadors and join in the performance if they share aspects of identity with them, though it may be the act of performing that is most powerful. Pupils and ambassadors perform their identities through small talk and physical interactions. Pupils enjoy this performance and the relationships that they forge with ambassadors in the social space that the performance allows. I suggest that, through participating in this performance, pupils are in a process of 'subjection' (Butler, 1997b). This performance then may be part of a process of naming and making future university students (Youdell, 2006).

However, as Butler (1997b: 83–4) suggests, there is a 'restriction in production' and this is an issue for the WP aims of these projects. The participation of black and working-class pupils from south east London schools at activities at Bankside may reinforce existing patterns of HE participation. Bankside provides a comfortable environment where pupils are surrounded by people like themselves. This process may actually serve to preserve the status quo rather than challenge it; likewise the attendance of white middle-class pupils at the G&T summer school at Royal. The relationships may also be undermined by the constraints imposed by regulations associated with child protection. Physical contact and responding to requests to join social networking sites like Facebook, for example, are expected aspects of developing social relationships among young people. The ambassadors' negative responses to pupils' requests in this context may consequently be experienced as rejection, damaging the fragile development of subject and HE identities that ambassadors have supported.

There are glimpses in this study of how carefully designed experiential activities that utilize ambassadors may actually challenge patterns of HE participation within subject areas. Ambassadors provide a challenge to assumptions about ethnic and gendered science and engineering identities (Walker, 2001) and pupils briefly take up these new 'ways of being' (Davies, 2006: 430). This type of ambassador work has potential to 'disrupt dominant discourses around science' and 'the identity of the scientist and interrupt dominant identity patters of (dis)identification' (Archer *et al.*, 2010: 21).

Pitfalls and potential

Student ambassador schemes are embedded within dominant contemporary neo-liberal discourses currently operating as regimes of truth within the HE system as a whole. Organizers and ambassadors work in institutions where discourses of the marketplace and of individualized choosers operating within the HE system are dominant. Within these institutions, ambassadors are positioned as marketers of their universities, and pupils as consumers. This has happened almost accidentally and without the consent or knowledge of organizers. The marketization of the HE system has been a relentless and powerful process that has carried all with it. With the radical changes to HE funding, this neo-liberal discourse is entrenching. The BIS (2011: 5) report on Higher Education pronounces that 'putting financial power into the hands of learners makes student choice meaningful'. Pupils are now fully fledged consumers in the HE marketplace. With the withdrawal of funds for the Connexions services and Aimhigher schemes, however, the opacity of this marketplace for young people is likely to increase.

The dominance of these neo-liberal discourses and the stratified positioning of HE institutions as businesses in competition for (different) pupils, has already pushed aside learning and teaching in WP work. Aspiration-raising is practised and understood by ambassadors in this study as marketing their institutions and their courses. Ambassadors enacting role models is similarly taken to imply promoting themselves, their institutions and their courses. The aim to widen participation through raising pupil aspirations positions pupils as lacking and idealizes middle-class aspirations. It is important to question whether this message benefits or undermines pupils and to question in this blizzard of marketing from HEIs what information pupils take on and whether this is likely to help them to make informed choices.

In the light of this study, the stringent targeting criteria (HEFCE, 2007) adopted by Aimhigher in an attempt to ensure that the most disadvantaged learners were reached are also revealed to be problematic for schemes using ambassadors. The Aimhigher Associates Scheme, for instance, positioned ambassadors similarly to teachers in schools to work with groups of disadvantaged learners in order to raise their aspirations. Despite assumptions that ambassadors are role models, this study suggests that positioning ambassadors in a context with many formal attributes, working with pupils who do not have established learner identities, is as likely to entrench a sense of difference as to encourage pupils to aspire to be like ambassadors. This approach is also problematic, as pupils are uncomfortable talking to peers about aspirations especially in relation to HE (Brooks, 2003a). Pupils', teachers' and ambassadors' responses to the scheme, outlined in the qualitative evaluation (HEFCE, 2010a), reflect these issues.

There are glimpses in this study of how ambassadors' interactions with pupils can contribute to challenging existing patterns of participation within universities and subject areas. Working collaboratively with ambassadors in learning contexts with a number of informal attributes can provide school pupils with an opportunity to enact student identities and to gain an insight into HE and their possible positioning within it. Pupils' perception of what is possible is currently defined by their peer group, family, school and social networks (Reay *et al.*, 2005; Brooks, 2003a); these interactions could extend pupils' perceptions of what 'constitutes a "feasible" choice' (Brooks, 2003a: 292). Interactions between ambassadors and pupils, however, need to be carefully planned to facilitate the development of relationships where such learning can take place. The information that ambassadors provide for pupils needs further interrogation, but I suggest that ambassadors can support and reinforce subject orientations and learning identities that pupils have already begun to develop at school.

There is potential to further develop the work undertaken by student ambassadors to more effectively disrupt existing gendered, raced and classed subject identities and patterns of participation within subject areas. There is also the potential to use activities to challenge credentialist forms of learning and to encourage pupils to consider subjects in the light of real-world applications. Student ambassadors could be a valuable resource within STEM education to challenge existing perceptions and STEM identities and promote equality in HE in these subject areas. This study illustrates the impact working with ambassadors collaboratively in 'safe learning spaces' (Burke, 2012: 187) can have on pupils, consolidating their subject interests in medicine and engineering and related subjects and reinforcing HE identities.

Appendix

Policies and projects involved in the research

There were a number of particular policies and projects that provided the financial support for WP work in the two universities during this study. Aimhigher provided funds to both institutions. Both institutions also drew funds from other sources for WP projects generally and specifically for STEM subject areas. At the time of the study, Bankside was the host university for a three-year project: the Accessing Engineering Project (AEP). Royal had a Medical Access Scheme (MAS) that has run at the university for eight years. The description of the policies outlines the context at the time of data collection, though these contexts have now changed dramatically.

Aimhigher

Aimhigher was one policy focusing on 'raising' the perceived 'low aspirations' of groups 'under-represented' in HE in England. Until 2004, this had two policy strands: Aimhigher Excellence Challenge and Partnership for Progression. In 2004, these strands were integrated to form Aimhigher. Aimhigher was funded by HEFCE and consisted predominantly of local 'partnerships' between HE institutions, further education colleges and school representatives from Local Education Authorities (LEAs). In 2006/7, Aimhigher had a budget of £85 million, which made it 'one of the largest initiatives for widening participation in England' (Hatt *et al.*, 2007).

The target groups for Aimhigher activities were delineated quite specifically in a HEFCE (2007: cover) document providing 'guidance on effective ways to target outreach activities at people from communities under-represented in higher education'. This document identified three stages of targeting: 'area level', 'learner level' and finally 'monitoring the effectiveness of targeting procedures' (10). This emphasis on monitoring was also present in the *Guidance for Aimhigher Partnerships* (HEFCE, 2008), which demanded a coherent evaluation plan from all area partnerships. Aimhigher partners were asked to identify specific students from the target groups to be selected for activities and to find out detailed information about these learners' backgrounds. The target groups identified in the guidance for Aimhigher partnerships (HEFCE, 2008: 4) were:

- people from lower socio-economic groups

- people from disadvantaged socio-economic groups who live in areas of relative deprivation where participation in HE is low
- 'looked after' children in the care system
- people with a disability or a specific learning difficulty.

The HEFCE (2007) document on effective targeting provided some detailed guidance on how to target groups. Whereas disabled people were to be targeted because of their disability, ethnic minority groups were not to be targeted because of their ethnicity. However, the guide explained that ethnic minority groups often 'live in the most disadvantaged communities and will therefore often form part of the key target group' (9). It was also noted that ethnic groups are currently over-represented in 'certain institutions and subjects' and 'so there are important issues of fair access' (8). The target age of students for Aimhigher activities was 13–30, to connect to the government's 50 per cent participation target, although again there was acknowledgement that it is important to work with younger children in order to 'sow the seeds of raised ambition'.

Student ambassadors were a ubiquitous feature of Aimhigher programmes and were used in a variety of ways. However, funding from Aimhigher was divided between the different stakeholders and, within HEIs, it was combined with other sources to fund WP activities. At Bankside specifically, posts within the WP unit were permanent Bankside posts, but were funded by a combination of resources drawn from different streams. There was a WP premium, which was allocated to Bankside by HEFCE on the basis of the postcodes of the student intake. HEFCE also provided match funding for the summer school. There was a summer school of smaller scale that was funded specifically by Aimhigher and administered by the local partnership. Another source of money was drawn from the TDA to fund the Student Associates Scheme. Although the WP unit at Bankside did function independently of Aimhigher, Aimhigher had a unifying influence over projects run by this and the other HEIs in the partnership.

At Royal, the WP unit relied more heavily on funding from Aimhigher, though again the WP officers had permanent posts (but have now been made redundant). As well as Aimhigher funding, the WP unit received a limited amount of money from the university to run Gifted and Talented (G&T) projects.

In many instances, the different HEIs worked on the same projects and in the same schools and colleges. Student ambassadors had Aspire T-shirts, which they wore during events, funded by Aspire, so that ambassadors from different institutions all wore the same uniform and were not easily identified as being from separate institutions.

STEM

The Accessing Engineering Project (AEP)

During 2008/9, Bankside was the lead institution for an HEFCE-funded project aiming to widen and increase participation in engineering – the AEP. While funding for this project did not go directly to the WP unit, fieldworkers appointed to the AEP worked alongside WP staff, and student ambassadors from Bankside were used within various parts of the project. The AEP had a similar agenda to other government WP policy. Project work was run in schools with under-represented groups to raise aspiration, though the ambition of the AEP was to particularly raise aspiration and awareness of futures in engineering and STEM more generally. The target groups were different to those identified for Aimhigher, as the groups under-represented in engineering HE courses differ to those under-represented in HE more generally. Girls and minority-ethnic groups were particularly identified as part of those targeted by the project's strong WP agenda. The project was not subject to the increasingly precise prescriptions from HEFCE (2007) about targeted groups and more broadly targeted anyone with no family history of HE.

Medical Access Scheme (MAS)

The MAS outreach ran separately to the WP unit at Royal, but had a clear WP agenda. The outreach programme for the MAS was described in a leaflet for the Royal School of Medicine as being a series of activities for students in Years 8–11. The aim of the programme was to encourage and enable young people from local boroughs to study at medical school. A series of activities were run over the course of the academic year, including placing medical students in the classroom over a four-week period, interactive half days introducing students to the work of medical professionals and the medical activity days which focus on a scientific or medical topic. This scheme had run for eight years; the ambition of the scheme was to recruit pupils from local boroughs to the MAS running at Royal. This degree course was six years in duration as opposed to the five years of a traditional medical degree. The additional year was designed to allow students who have not started the course with the same achievements at A-level, but have the 'academic potential for a career in medicine' to catch up with their peers on a traditional medical degree. In 2008/09 (when the data was collected for this study), only students from non-selective state schools from 16 local boroughs were entitled to apply, but this was relaxed in 2009/10 to include all London boroughs.

The outreach programme is testimony to the long history of WP work at Royal. The programme received funds from a local charity which celebrated its 700-year anniversary in 2008 and raised funds during this anniversary year for outreach activities, text books and prizes for the MAS. This charity was the pre-eminent source of funding for the outreach programme. Aimhigher provided an additional source of funding for some activities.

While medical students on the MAS were involved in outreach activities for a number of years, this was formalized into an ambassador programme based on the one running within the WP unit at Royal. A pool of 21 student ambassadors supported activities on the MAS in 2008/9 and this number increased in 2009/10 to approximately 50 ambassadors.

References

Abbot, D.A, Meridith W.H., Self-Kelly, R. and Davis, M.E (1997) 'The influence of a Big Brothers program on the adjustment of boys in single parent families'. *Journal of Psychology*, 131, 143–56.

Albanese, M. and Mitchell, S. (1993) 'Problem-based learning: A review of the literature on its outcomes and implementation issues'. *Academic Medicine*, 68 (1), 52–81.

Allen, I. (1988) *Doctors and Their Careers*. London: Policy Studies Institute.

Allen, T.D. and Eby, L.T. (2007) *The Blackwell Handbook of Mentoring: A multiple perspectives approach*. Oxford: Blackwell Publishing.

Ambrose, S.A., Kristin, L., Dunkle, B., Lazarus, B., Nair, I. and Harkus, D.A. (1997) *Journeys of Women in Science and Engineering*. Philadelphia: Temple University Press.

Anderson, N. (2003) *Discursive Analytic Strategies: Understanding Foucault, Koselleck, Laclau, Luhmann*. Bristol: Policy Press.

Andrews, J. and Clark, R. (2011) *Peer Mentoring Works! How peer mentoring enhances student success in higher education*. Birmingham: Aston University. Online. http://www.heacademy.ac.uk/assets/documents/what-works-student-retention/Aston_What_Works_Final_Report.pdf [16/03/2014].

Andrews, J., Clark, R. and Thomas, L. (eds) (2012) *Compendium of Effective Practice in Higher Education Retention and Success*. Birmingham and York: Aston University and the Higher Education Academy. Online. http://www.heacademy.ac.uk/assets/documents/what-works-student-retention/What_Works_Compendium_Effective_Practice.pdf [16/03/2014].

Archer, L., Dewitt, J., Osborne, J., Dillon, J., Willis, B. and Wong, B. (2010) '"Doing" science versus "being" a scientist: Examining 10/11-year-old schoolchildren's constructions of science through the lens of identity'. *Science Education*, 94 (4), 617–39.

— (2012) 'Science aspirations, capital and family habitus: How families shape children's engagement and identification with science'. *American Education Research Journal*, 49 (5), 881–908.

Archer, L., Hutchings, M. and Ross, A. (eds) (2003) *Higher Education and Social Class: Issues of exclusion and inclusion*. London: RoutledgeFalmer.

Archer, L. and Yamashita, H. (2003) 'Knowing their limits? Identities, inequalities and inner city school leavers post 16 aspirations'. *Journal of Education Policy*, 18 (1), 53–69.

Arlett, C., Lamb, F., Dales, R., Willis, L. and Hurdle, E. (2010) 'Meeting the needs of industry: The drivers for change in engineering education'. *Engineering Education*, 5 (2), 18–25.

Arnot, M. and Reay, D. (2007) 'A sociology of pedagogic voice: Power, inequality and pupil consultation'. *Discourse: Studies in the Cultural Politics of Education*, 28 (3), 311–25.

Aseltine, R.H., Dupre, M. and Lamlein, P. (2000) 'Mentoring as a drug prevention strategy: An evaluation of Across the Ages'. *Adolescent and Family Health*, 1 (1), 11–20.

References

Austin, M. and Hatt, S. (2005) 'The messengers are the message: A study of the effects of employing Higher Education student ambassadors to work with school students'. *Widening Participation and Lifelong Learning*, 7 (1), 22–9.

Baker, D., Krouse, S., Yasar, S., Roberts, C. and Robinson-Kurpius, S. (2007) 'An intervention to address gender issues on a course of design, engineering and technology for science educators'. *Journal of Engineering Education*, 96 (3), 213–26.

Ball, S.J. (1994) *Education Reform: A critical and post-structuralist approach*. Buckingham: Open University Press.

— (2003) *Class Strategies and the Education Market: The middle classes and social advantage*. London: Routledge.

Ball, S.J., Maguire, M. and Macrae, S. (2000) *Choice, Pathways and Transitions Post-16: New youth, new economies in the global city*. London: Falmer.

Ball, S.J. and Vincent, C. (1998) '"I heard it on the grapevine": "Hot" knowledge and school choice'. *British Journal of Sociology of Education*, 19 (3), 377–400.

Beck, U. (1992) *Risk Society: Towards a new modernity*. Newbury Park, CA: Sage.

Beck, V., Fuller, A. and Unwin, L. (2006) 'Safety in stereotypes? The impact of gender and "race" on young people's perceptions of their post-compulsory education and labour market opportunities'. *British Educational Research Journal*, 32 (5), 667–86.

Beckett, D. and Hager, P. (2002) *Life, Work And Learning: Practice in postmodernity*. London: Routledge.

Benwell, B. and Stokoe, E. (2006) *Discourse and Identity*. Edinburgh: Edinburgh University Press Ltd.

Billig, M. (1991) *Ideology and Opinions: Studies in rhetorical psychology*. London: Sage.

BIS (2011) *Higher Education: Students at the heart of the system*. London: The Stationery Office. Online. http://webarchive.nationalarchives.gov.uk/+/http://discuss.bis.gov.uk/hereform/white-paper/ [13/05/2014].

Blanden, J. and Machin, S. (2004) 'Educational inequality and the expansion of UK higher education'. *Scottish Journal of Political Economy*, 51 (2), 230–49.

Blanton, H., Crocker, J. and Miller, D.T. (2000) 'The effects of in-group versus out-group social comparison on self-esteem in the context of a negative stereotype'. *Journal of Personality and Social Psychology*, 51, 1173–82.

Blinn-Pike, L. (2007) 'The benefits associated with youth mentoring relationships'. In Allen, T.D. and Eby L.T. (eds) *The Blackwell Handbook of Mentoring: A multiple perspectives approach*. Oxford: Blackwell Publishing.

Boud, D., Cohen, R. and Walker, D. (eds) (1993) *Using Experience for Learning*. Buckingham: SRHE and Open University Press.

Boud, D. and Miller N. (eds) (1996) *Working with Experience: Animated learning*. London: Routledge.

Boursicot, K. and Roberts, T. (2009) 'Widening participation in medical education: Challenging elitism and exclusion'. *Higher Education Policy*, 22 (1), 19–36.

Bowler, L. (2004) 'Ethnic profile of the doctors in the United Kingdom'. *British Medical Journal*, 329 (7466), 583–4.

Bridges, D. (2005) 'Widening participation in higher education: "The philosopher and the bricklayer" revisited'. Paper presented at the Annual Conference of the Philosophy of Education Society of Great Britain, Oxford, 1–3 April.

Brooks, R. (2003a) 'Young people's higher education choices: The role of family and friends'. *British Journal of Sociology of Education*, 24 (3), 283–98.

— (2003b) 'Discussing higher education choices: Differences and difficulties'. *Research Papers in Education*, 18 (3), 237–58.

— (2004) '"My mum would be as pleased as punch if I actually went, but my dad seems a bit more particular about it": paternal involvement in young people's higher education choices'. *British Educational Research Journal*, 30 (4), 495–514.

Broughton, N. (2013) *In the Balance: The STEM human capital crunch*. London: Social Market Foundation.

Brown, G. (2008) 'We'll use our schools to break down class barriers'. *The Observer*, 10 February.

Brown, P. and Hesketh, A. (2004) *The Mismanagement of Talent: Employability and jobs in the knowledge economy*. Oxford: Oxford University Press.

Brown, P., Lauder, H. and Ashton, D. (2008) *Education, Globalisation and the Knowledge Economy: A commentary by the Teaching and Learning Research Programme*. London: TLRP Institute of Education.

Burke, P.J. (2002) *Accessing Education: Effectively widening participation*. Stoke-on-Trent: Trentham.

— (2006) 'Fair Access? Exploring gender, access and participation beyond entry to higher education'. In Leathwood, C. and Francis, B. (eds) *Gender and Lifelong Learning: Critical feminist engagements*. London: Routledge.

— (2012) *The Right to Higher Education: Beyond widening participation*. London: Routledge.

Butler, J. (1990) *Gender Trouble: Feminism and the subversion of identity*. London: Routledge.

— (1997a) *Excitable Speech: A politics of the performative*. London: Routlege.

— (1997b) *The Psychic Life of Power: Theories in subjection*. Stanford: Stanford University Press.

— (2004) *Precarious Life: The powers of mourning and violence*. London: Verso.

Cabinet Office (2008) *Getting On, Getting Ahead: A discussion paper analysing the trends and drivers of social mobility*. Online. http://dera.ioe.ac.uk/8835/1/gettingon.pdf [15/03/2014].

Canavan, B., Magill, J. and Love, D. (2002) 'A study of factors affecting perception of science, engineering and technology (SET) in young people'. Paper presented at the International Conference on Engineering Education, Manchester, August.

Cano, C., Kimmel, H., Koppel, N. and Muldrow, D. (2001) 'A step for women into the engineering pipeline'. Paper presented at the ASEE/IEEE Frontiers in Education Conference, Reno, October.

Carpenter, C. and Kerrigan, M. (2009) *What's the Score? An evaluation of the Aimhigher Boys into Higher Education Using Football project*. Leicester: Aimhigher Leicester City and Leicestershire.

Catalano, R.F., Hawkins, J.D., Berglund, L.M., Pollard, J.A. and Arthur, M.W. (2002) 'Prevention science and positive youth development: Competitive or cooperative frameworks?' *Journal of Adolescent Health*, 31, 230–9.

Cave, G. and Quint, J. (1990) *Career Beginnings Impact Evaluation*. New York: Manpower Demonstration and Research Corporation.

References

Cavell, T.A., Meehan, B.T., Heffer, R.W. and Holladay, J.J. (2002) 'The natural mentors of adolescent children of alcoholics (COAs): Implications for preventative practices'. *Journal of Primary Prevention*, 23 (1), 23–42.

CBI (Confederation of British Industry) (2010) *Ready to Grow: Business priorities for education and skills, education and skills survey*. Online. http://www.britishcouncil.org/zh/education_and_skills_survey_2010.pdf [16/03/2014].

Chapman, B. and Ryan, C. (2003) 'The access implications of income contingent charges for higher education: Lessons from Australia'. In *CEPR Discussion Papers*, 436. Canberra: Centre for Economic Policy Research.

Charmaz, K. (2003) 'Grounded theory: Objectivist and constructivist methods'. In Denzin N.K. and Lincoln, Y.S. (eds) *Strategies of Qualitative Inquiry*. London: Sage.

Chen, C., Greenberger, E., Farruggia, S., Bush, K. and Dong, Q. (2003) 'Beyond parents and peers: The role of very important non-parental adults (VIPs) in adolescent development in China and the United States'. *Psychology in Schools*, 40 (1), 35–50.

Chesler, N.C. and Chesler, M.A. (2002) 'Gender-informed mentoring strategies for women engineering scholars: On establishing a caring community'. *Journal of Engineering Education*, 91 (1), 49–55.

Chilosi, D. (2008) 'Employability Paper'. Greenwich University.

Chisholm, L., Hoskins, B. and Glahn, C. (eds) (2005) *Trading Up: Potential and performance in non-formal learning*. Strasbourg: Council of Europe.

Church, E. and Kerrigan, M. (2010) *An Evaluation of the Aimhigher Northamptonshire Associates Programme*. Loughborough: Aimhigher in the East Midlands. Online. http://www.heacademy.ac.uk/assets/documents/aim_higher/AHNorthants_Associates_Programme.pdf [16/03/2014].

Clutterbuck, D. and Ragins, B.R. (2002) *Mentoring and Diversity: An International Perspective*. Oxford: Butterworth-Heinemann.

Coffield, F. (2000) *The Necessity of Informal Learning*. Bristol: Policy Press.

Cohen, S. and Willis T.A. (1985) 'Stress, social support, and the buffering hypothesis'. *Psychological Bulletin*, 98 (2), 310–57.

Colley, H. (2003) 'Engagement mentoring for "disaffected" youth: A new model of mentoring for social inclusion'. *British Educational Research Journal*, 29 (4), 521–32.

— (2005) 'Formal and informal models of mentoring for young people: Issues for democratic and emancipatory practice'. In Chisholm, L., Hoskins, B. and Glahn, C. (eds) *Trading up: Potential and performance in non-formal learning*. Strasbourg: Council of Europe.

Colley, H., Hodkinson, P. and Malcolm, J. (2003) *Informality and Formality in Learning: A report for the Learning and Skills Research Centre*. London: Learning and Skills Research Centre. Online. http://lllp.iugaza.edu.ps/Files_Uploads/634791628087049086.pdf [17/03/2014].

Collins, R.L. (1996) 'For better or worse: The impact of upward social comparisons on self-evaluation'. *Psychological Bulletin*, 119 (1), 51–69.

Connor, H., Tyers, C., Modood, T. and Hillage, J. (2004) *Why the Difference? A closer look at higher education minority ethnic students and graduates*. Research Report 552. Nottingham: DfES.

Crenshaw, K. (1989) 'Demarginalizing the intersection of race and sex: A black feminist critique of antidiscrimination doctrine, feminist theory and antiracist politics'. *University of Chicago Legal Forum*, 139–67.

Crozier, G., Reay, T. and Clayton, J. (2010) 'The socio-cultural and learning experiences of working-class students in higher education'. In David, M. *et al.* (ed.) *Improving Learning by Widening Participation in Higher Education.* London: Routledge.

Cunningham, C. and Knight, M. (2004) 'Draw an engineer test: Development of a tool to investigate students' ideas about engineers and engineering'. Proceedings of the 2004 American Society for Engineering Education Annual Conference and Exposition, Portland, OR, 20–23 June. Washington, DC: ASEE.

Cunningham, C., Lachapelle, C. and Lindgren-Streicher, A. (2005) 'Assessing elementary school students' conceptions of engineering and technology'. Proceedings of the 2005 American Society for Engineering Education Annual Conference and Exposition, Portland, OR, 12–15 June. Washington, DC: ASEE.

— (2006) 'Elementary teachers' understandings of engineering and technology'. Proceedings of the 2006 American Society for Engineering Education Annual Conference and Exposition, Portland, OR, 18–21 June. Washington, DC: ASEE.

David, M., Bathmaker, A., Crozier, G., Davis, P., Hockings, C., Parry, G., Reay, D., Vignoles, A. and Williams, J. (2010) *Improving Learning by Widening Participation in Higher Education.* London: Routledge.

David, M., Coffey, A., Connolly, P., Nayak, A. and Reay, D. (2006) 'Troubling identities: Reflections on Judith Butler's philosophy for the sociology of education'. *British Journal of Sociology of Education*, 27 (4), 421–4.

Davies, B. (2006) 'Subjectification: The relevance of Butler's analysis for Education'. *British Journal of Sociology of Education*, 27 (4), 425–38.

Davies, B. and Harre, R. (1990) 'Positioning: The discursive production of selves'. *Journal for the Theory of Social Behaviour*, 20 (1), 44–63.

Davis, J.E. (2001) 'Transgressing the masculine: African American boys and the failure of schools'. In Wayne, M. and Meyenn, B. (eds) *What About the Boys? Issues of masculinity in school.* Buckingham: Open University Press.

Delgado, R. (1991) 'Affirmative action as a majoritarian device: Or, do you really want to be a role model?' *Michigan Law Review*, 89 (5), 1222–32.

DIUS (Department for Innovation, Universities and Skills) (2008) *Innovation Nation White Paper.* Online. http://webarchive.nationalarchives.gov.uk/+/http://bis.gov.uk/policies/innovation/white-paper [17/03/2014].

Doyle, M. and Griffin, M. (2012) 'Raised aspirations and attainment? A review of the impact of Aimhigher (2004–2011) on widening participation in higher education in England'. *London Review of Education*, 10 (1), 75–88.

Draut, M. (2000) 'Non-formal learning: Implicit learning and tacit knowledge'. In Coffield, F. (ed.) *The Necessity of informal Learning.* Bristol: Policy Press.

DuBois, D.L., Hollway, B.E., Valentine, J.C. and Cooper, H. (2002) 'Effectiveness of mentoring programs for youth: A meta analytic review'. *American Journal of Community Psychology*, 30 (2), 157–96.

Edley, N. (2001) 'Analysing masculinity: Interpretative repertoires, ideological dilemmas and subject positions'. In Wetherell, M., Taylor, S. and Yates, S.J. (eds) *Discourse as Data: A guide for analysis.* Milton Keynes: Open University Press.

References

EEA (Engineering Education Alliance Report) (2002) *Attracting More Entrants into the World of Engineering.* Online. http://tinyurl.com/qbmth3g [17/03/2014].

Eisenhart, M., Finkel, E. and Marion, S. (1996) 'Creating the conditions for scientific literacy: A reconsideration'. *American Educational Research Journal,* 33 (2), 261–95.

Engineering UK (2011) *The State of Engineering.* Online. http://tinyurl.com/p27f3pf [17/03/2014].

Ensher, E.A., Grant-Vallone, E.J. and Marelich, W.D. (2002) 'Effects of perceived attitudinal and demographic similarity on proteges' support and satisfaction gained from their mentoring relationships'. *Journal of Applied Psychology,* 32, 1407–30.

Ensher, E.A. and Murphy, S.E. (1997) 'Effects of race, gender, perceived similarity and contact on mentor relationships'. *Journal of Vocational Behaviour,* 57, 1–21.

Eraut, M. (2000) 'Non-formal learning, implicit learning and tacit knowledge'. In Coffield, F. (ed.) (2000) *The Necessity of Informal Learning.* Bristol: Policy Press.

Etkowit, H.C., Kemelgor, C. and Uzzi, B. (2000) *Athena Unbound: The advancement of women in science and technology.* New York: Cambridge University Press.

European Commission (2001) *Making a European Area of Lifelong Learning a Reality.* Online. http://eur-lex.europa.eu/LexUriServ/LexUriServ.do?uri=COM:2001:0678:FIN:EN:PDF [17/03/2014].

Evans, K. (2006) 'Achieving equity through "gender autonomy": Challenges for vocational education and training (VET) policy and practice'. *Journal of Vocational Education and Training,* 58, 393–408.

— (2007) 'Concepts of bounded agency in education, work and personal lives of young adults'. *International Journal of Psychology,* 42 (2), 1–9.

Evans, K., Hodkinson, P., Rainbird, L., Unwin, L. (2006) *Improving Workplace Learning.* London: Routledge.

Fielding, M. (2004) 'Transformative approaches to student voice: Theoretical underpinnings, recalcitrant realities'. *British Educational Research Journal,* 30 (2), 295–311.

Finch, J. (1987) 'The vignette technique in survey research'. *Sociology,* 21 (1), 105–14.

Fordham, S. (1996) *Blacked Out: Dilemmas of race, identity and success at Capital High.* Chicago: University of Chicago Press.

Foucault, M. (1977) *Discipline and Punish.* London: Tavistock.

— (1980) *Power/Knowledge.* Brighton: Harvester.

— (1998) *The History of Sexuality: The will to knowledge.* London: Penguin.

Francis, B. (2002) 'Is the future really female? The impact and implications of gender for 14–16 year olds' career choices'. *Journal of Education and Work,* 15 (1), 75–88.

Furlong, A. and Cartmel. E. (1997) *Young People and Social Change: Individualization and risk in later modernity.* Milton Keynes: Open University Press.

Gale, T. and Tranter, D. (2011) 'Social justice in Australian higher education policy: An historical and conceptual account of student participation'. *Critical Studies in Education,* 52 (1), 29–46.

Galindo-Rueda, F., Vignoles, A. and Marcenaro-Gutierrez, O. (2004) *The Widening Socio-economic Gap in UK Higher Education*. London: Centre for the Economics of Education, London School of Economics.

Gartland, C. (2009) 'The London Engineering Project: long term outcomes'. In Royal Academy of Engineering *London Engineering Project Evaluation Report*. Online. http://www.thelep.org.uk/national/lepreport/projectoutputsandoutcomes [17/03/2014].

— (2012/13) 'Marketing Participation? Student ambassadors' contribution to widening participation schemes in engineering and medicine at two contrasting universities'. *Journal of Widening Participation and Lifelong Learning*, 12 (3), 102–119.

— (2014) 'Student ambassadors: "Role models", learning practices and identities'. *British Journal of Sociology of Education*. Online. http://www.tandfonline.com/doi/full/10.1080/01425692.2014.886940#tabModule [17/03/2014].

Gartland, C. Hawthorn, H. and McLoughlin, C. (2010) 'Discourses, identities and learning: Implications for the training of student ambassadors in engineering'. Paper presented at the Engineering Education Conference on *Inspiring the Next Generation of Engineers*, Birmingham, July.

Gartland, C. and Paczuska, A. (2007) 'Student ambassadors, trust and HE choices'. *Journal of Access Policy and Practice*, 4 (2), 108–33.

Gaskell, G. (2000) 'Individual and group interviewing'. In Bauer, M. and Gaskell G. (eds) *Qualitative Researching with Text, Image and Sound*. London: Sage.

Gerson, K. and Horowitz, R. (2003) 'Observation and interviewing: Options and choices in qualitative research'. In May, T. (Ed.) *Qualitative Research in Action*. London: Sage.

Gillborn, D. and Youdell, D. (2000) *Rationing Education: Policy, practice, reform and equity*. Buckingham: Open University Press.

Goldacre, M.J., Davidson, J.M. and Lambert, T.W. (2004) 'Country of training and ethnic origin of UK doctors: Database and survey studies'. *British Medical Journal*, 329 (7466), 597.

Gorard, S., Adnett, N., May, H., Slack, K., Smith, E. and Thomas, L. (2007) *Overcoming the Barriers to Higher Education*. Stoke-on-Trent: Trentham Books.

Gosa, T. and Young, H. (2006) 'The construction of oppositional culture in hip-hop music: An in-depth case analysis of Kanye West and Tupac Shakur. Baltimore, MD: Johns Hopkins University.

Grant, J., Jones, H. and Lambert, T. (2002) *An Analysis of Trends in Applications to Medical School*. Milton Keynes: Open University Centre for Education in Medicine/UK Medical Careers Research Group.

Greenhalgh, T., Seyan, K. and Boynton, P. (2004) '"Not a university type": Focus group study of social class, ethnic, and sex differences in school pupils' perceptions about medical school'. *British Medical Journal*, 328 (7455), 1541.

Greenwood, C., Harrison, M. and Vignoles, A. (Department of Quantitative Social Science, Institute of Education) (2011) *The Labour Market Value of STEM qualification and Occupations*: *An analysis for the Royal Academy of Engineering*. Online. http://www.raeng.co.uk/news/publications/list/reports/STEM_WageReturns.pdf [17/03/2014].

References

Grossman, J.B. and Rhodes, J.E. (2002) 'The test of time: Predictors and effects of duration in youth mentoring'. *American Journal of Community Psychology*, 30, 199–219.

Hackett, G. (1985) 'Role of mathematics self-efficacy in the choice of math-related majors of college women and men: A path analysis'. *Journal of Counseling Psychology*, 32, 47–56.

Hager, P. (2005) 'Current theories of workplace learning: A critical assessment'. In Bascia, N., Cumming A., Datnow A., Leithwood, K. and Livingstone, D. (eds) *International Handbook of Educational Policy Part 2*. Dordrecht: Springer.

Hager, P. and Hodkinson, P. (2011) 'Becoming as an appropriate metaphor for understanding professional learning'. In Scanlon, L. (ed.) *"Becoming" A Professional: An interdisciplinary analysis of professional learning*. Dordrecht: Springer.

Hall, S. (1997) 'Foucault: Power, knowledge and discourse'. In Wetherell, M., Taylor, S. and Yates, S.J. (eds) (2001) *Discourse Theory and Practice: A reader*. London: Sage.

Harding, S. (1991) *Whose Science? Whose Knowledge?: Thinking from women's lives*. Milton Keynes: Open University Press.

Harre, R. and Van Langenhove, L. (eds) (1999) *Positioning Theory*. Oxford: Blackwell.

Harrison, M. (2011) 'Subjects and skills: Issues in post-14 STEM education'. Visiting professor inaugural lecture at the Institute of Education, London, 14 April.

Hatt, S., Baxter, A. and Tate, J. (2007) 'From Policy to Practice: pupil's responses to WP initiatives'. *HE Quarterly*, 61 (3), 266–83.

— (2009) '"It was definitely a turning point!" A review of Aimhigher summer schools in the south west of England'. *Journal of Further and Higher Education*, 33 (4), 333–46.

HEA (2009–10) *Handbook for Aimhigher Associates*. Online. http://www.heacademy.ac.uk/assets/documents/aimhigherassociates/Associates-Handbook-2009-10.pdf.

Heath, S., Brooks, R., Cleaver, E. and Ireland, E. (2009) *Researching Young People's Lives*. London: Sage.

HEFCE (2005) *Evaluation of Aimhigher: Excellence Challenge Interim Report*. Online. https://www.education.gov.uk/publications/eOrderingDownload/RB648.pdf [17/03/2014].

HEFCE (2007) *Higher Education Outreach: Targeting disadvantaged learners*. Online. http://www.hefce.ac.uk/pubs/hefce/2007/07_12/07_12.pdf [17/03/2014].

HEFCE (2008) *Guidance for Aimhigher Partnerships: Updated for the 2008–2011 programme*. Online. http://www.hefce.ac.uk/media/hefce1/pubs/hefce/2008/0805/08_05.pdf [13/05/2014].

HEFCE (2009a) *Aimhigher Associates Scheme: Guidance and planning for the national phase, 2009–2011*. Online. http://dera.ioe.ac.uk/147/1/AH_135435454321.pdf [17/03/2014].

HEFCE (2009b) *Strategically Important and Vulnerable Subjects: The HEFCE advisory group's 2009 report*. Online. http://www.hefce.ac.uk/pubs/hefce/2010/10_09/10_09_354235456.pdf [17/03/2014].

HEFCE (2010a) *Qualitative Evaluation of the Aimhigher Associates Programme Pathfinder*. Online. http://www.hefce.ac.uk/pubs/rereports/year/2010/evalaimhigherpathf/ [17/03/2014].

HEFCE (2010b) *Trends in Young Participation in Higher Education: Core results for England*. Online. http://www.hefce.ac.uk/pubs/hefce/2010/10_03/10_03.pdf [17/03/2014].

HEFCE (2011) *Aimhigher Associates Scheme: Patterns of participation during the first year*. Online. http://www.hefce.ac.uk/pubs/year/2011/201135/. [17/03/2014].

HESA (2008/09) *Student population*. Online. https://www.hesa.ac.uk/intros/stu0809 [11/06/2014].

HESA (2009/10) *Student population*. Online. http://www.hesa.ac.uk/index.php/content/view/1974/278/ [07/2011].

Hey, V. (1997) *The Company She Keeps: An ethnography of girls' friendship*. Buckingham: Open University Press.

— (2006) 'The politics of performative resignification: translating Judith Butler's theoretical discourse and its potential for a sociology of education'. *British Journal of Sociology of Education*, 27 (4), 439–57.

Hey, V., Creese, A., Daniels, H., Fielding, S. and Leonard, D. (2001) '"Sad, bad or sexy boys": Girls' talk in and out of the classroom'. In Wayne, M. and Meyenn, B. (2001) *What About the Boys? Issues of masculinity in school*. Buckingham: Open University Press.

Hey, V. and Leathwood, C. (2009) 'Passionate attachments: Higher education, policy, knowledge, emotion and social justice. *Higher Education Policy*, 22, 101–18.

Hockings, C., Cooke, S. and Bowl, M. (2010) 'Learning and teaching for social diversity in higher education'. In David, M. *et al.* (eds) *Improving Learning by Widening Participation in Higher Education*. London: Routledge.

Hodkinson, P. (2004) 'Research as a form of work: Expertise, community and methodological objectivity'. *British Educational Research Journal*, 30 (1), 9–26.

Hodkinson, P. and Hodkinson, H. (2001) 'Problems of measuring learning and attainment in the workplace: Complexity, reflexivity and the localized nature of understanding'. Paper presented at the conference *Context, Power and Perspective: Confronting challenges to improving attainment in learning at work*, University College of Northampton, 8–10 November.

Hodkinson, P. and Macleod, F. (2007) 'Contrasting concepts of learning and contrasting research methodologies'. Paper presented at the Educational Research and TLRP Annual Conference, Cardiff, 26–27 November.

Hodkinson, P., Sparkes, A. and Hodkinson, H. (1996) *Triumphs and Tears: Young people, markets and the transition from school to work*. London: David Fulton Publishers.

Hollway, W. (1984) 'Gender difference and the production of subjectivity'. In Henriques, J., Hollway, W., Urwin, C., Venn, C. and Walkerdine, V. (eds) *Changing the Subject*. London: Methuen.

House of Commons Education Committee (2013) *Careers guidance for young people: the impact of the new duty on schools*. Seventh Report of Session 2012–13. Vol. 1.

Hughes, G. (2000) 'Marginalization of socio scientific material in science–technology–society science curricula: Some implications for gender inclusivity and curriculum reform'. *Journal of Research in Science Teaching*, 5, 426–40.

References

— (2001) 'Exploring the availability of student scientist identities within curriculum discourse: An anti-essentialist approach to gender-inclusive science'. *Gender and Education,* 13 (3), 275–90.

IMD (2004) *Index of Multiple Deprivation 2004.* Online. http://data.gov.uk/dataset/imd_2004 [17/03/2014].

Jackson, C.H., Kite, M.E. and Branscombe, N.R. (1996) 'African American women's mentoring experiences'. Paper presented at the annual meeting of the American Psychological Association, Toronto, Canada, August.

Johnson, A.W. (1997) 'Mentoring at-risk youth: A research review and evaluation of the impacts of the Sponsor-A-Scholar Program on student performance'. PhD. diss., University of Pennsylvania.

— (1999) *An Evaluation of the Long-Term Impact of Sponsor–a–Scholar (SAS) Program on Student Performance.* Princeton, NJ: Mathematical Policy Research, Inc.

Johnson, F. and Aries, E. (1983a) 'Conversational patterns among same-sex pairs of late-adolescent close friends'. *Journal of Genetic Psychology,* 142 (2), 225–38.

— (1983b) 'The talk of women friends'. *Women's Studies International Forum,* 6 (4), 353–61.

Jones, A. (2004) 'Social anxiety, sex, surveillance, and the "safe" teacher'. *British Journal of Sociology of Education,* 25 (1), 53–66.

Kahlenberg, R.D. (ed.) (2004) *America's Untapped Resource: Low-income students in higher education.* New York: Century Foundation Press.

Kalbfleish, P.J. and Davies, A. (1991) 'Minorities and mentoring: Managing the multi-cultural institution', *Communication Education,* 40, 266–271.

Keller, T.E. (2007) 'Youth mentoring: Theoretical and methodological issues'. In Allen, T.D. and Eby, L.T. (2007) *The Blackwell Handbook of Mentoring: A multiple perspectives approach.* Oxford: Blackwell Publishing.

Kelly, A. (1987) 'The construction of masculine science'. In Kelly, A. (ed.) *Science for Girls.* Milton Keynes: Open University Press.

Kerrigan, M. and Church, E. (2011) '"I can do that too if I work hard": A longitudinal study of Aimhigher Leicester City and Leicestershire learners'. Leicester: Aimhigher Leicester City and Leicestershire. Online. http://www.heacademy.ac.uk/assets/documents/aim_higher/CEP_report_Leicestershire_finalPDF.pdf [17/03/2014].

Kitwana, B. (2002) *The Hip-Hop Generation: Young blacks and the crisis in African American culture.* New York: Basic Civitas Books.

Klaw, E.L. and Rhodes, J.E. (1995) 'Mentor relationships and the career development of pregnant and parenting African-American teenagers'. *Psychology of Women Quarterly,* 19 (4), 551–62.

Kolb, D.A. (1984) *Experiential Learning: Experience as a source of learning and development.* New Jersey: Prentice-Hall.

Kvale, S. (1996) *InterViews: An introduction to qualitative research interviewing.* London: Sage.

Kyriacou, C. (2009) *Effective Teaching in Schools: Theory and practice.* Cheltenham: Nelson Thornes Ltd.

Lambert, R. (2003) *Lambert Review of Business–University Collaboration. Final report.* Norwich: HMSO. Online. www.eua.be/eua/jsp/en/upload/lambert_review_final_450.1151581102387.pdf [17/03/2014].

Lauder, H., Brown, P. and Ashton, D. (2008) 'Globalisation, skill formation and the varieties of capitalism approach'. *New Political Economy*, 13 (1), 19–35.

Lave, J. and Wenger, E. (1991) *Situated Learning: Legitimate peripheral participation*. Cambridge: Cambridge University Press.

Leathwood, C. (2004) 'A critique of institutional inequalities in higher education or an alternative to hypocrisy for higher educational policy'. *Theory and Research in Education*, 2 (1), 31–48.

Leathwood, C. and Francis, B. (eds) (2006) *Gender and Lifelong Learning: Critical feminist engagements*. Oxon: Routledge.

Leitch, S., Lord (2006) *Leitch Review of Skills. Prosperity for all in the global economy – world class skills. Final report*. Norwich: HMSO. Online. www.delni.gov.uk/leitch_finalreport051206[1]-2.pdf [17/03/2014].

Liang, B. and Grossman, J.M. (2007) 'Diversity and youth mentoring relationships'. In Allen, T.D. and Eby, L.T. *The Blackwell Handbook of Mentoring: A multiple perspectives approach*. Oxford: Blackwell Publishing.

Lightbody, P. and Durndell, A. (1996) 'Gendered career choice: Is sex-stereotyping the cause or the consequence?' *Educational Studies*, 22 (2), 133–46.

Little, A.J. and Leon de la Barra, A. (2009) 'Attracting girls to science, engineering and technology: An Australian perspective'. *European Journal of Engineering Education*, 34 (5), 439–45.

Liu, Y. (2013) 'Meritocracy and the Gaokao: A survey study of higher education selection and socio-economic participation in East China'. *British Journal of Sociology of Education*, 34 (5–6), 868–87.

Lockwood, P. and Kunda, Z. (1997) 'Superstars and me: Predicting the impact of role models on the self'. *Journal of Personality and Social Psychology*, 73 (1), 91–103.

Mac an Gaill, M. (1994) *The Making of Men*. Buckingham: Open University Press.

Machin, S. and Vignoles, A. (2004) 'Education inequality: The widening socio-economic gap'. *Fiscal Studies*, 107–28.

McManus, I.C. (2002) 'Medical school applications — A critical situation'. *British Medical Journal*, 325 (7368), 786–7.

McPartland, J.M. and Nettles, S.M. (1991) 'Using community adults as advocates or mentors for at-risk middle school students: A two year evaluation of Project RAISE'. *American Journal of Education*, 99, 568–86.

McRobbie, A. (2006) 'Four technologies of young womanhood'. Paper presented at TU-Berlin Zentrum für Interdisziplinäre Frauen- und Geschlecterforschung, Berlin, 31 October.

Major, B., Sciacchitano, A.M. and Crocker, J. (1993) 'In-group vs. out-group comparison and self-esteem'. *Personality and Social Psychology Bulletin*, 19 (6), 711–21.

Malcolm, J. (2000) 'Joining, invading, reconstructing: Participation for a change?' In Thompson, J. (2000) *Stretching the Academy*. Leicester: NIACE.

Marx, D.M. and Roman, J.S. (2002) 'Female role models: Protecting women's maths test performance'. *Personality and Social Psychology Bulletin*, 28 (9), 1183–93.

Mason, J. (1998) 'Qualitative researching'. In May, T. (2003) *Qualitative Research in Action*. London: Sage.

References

Maynard, M. (1998) 'Feminists knowledge and the knowledge of feminisms: Epistemology, theory, methodology and method'. In May, T. and Wiliams, M. (eds) *Knowing the Social World*. Buckingham: Open University Press.

Mirza, H. (2008) *Race, Gender and Educational Desire: Why black women succeed and fail*. London: Routledge.

Modood, T. (2004) 'Capitals, ethnic identity and educational qualifications'. *Cultural Trends*, 12 (2), 87–105.

Moreau, M.P. and Leathwood C. (2006) Graduates' employment and the discourse of employability: A critical analysis'. *Journal of Education and Work*, 19 (4), 305–24.

Morley, L. (2001) 'Producing New Workers: Quality, equality and employability in higher education'. *Quality in Higher Education*, 7 (2), 131–38.

— (2003) *Quality and Power in Higher Education*. Maidenhead: SRHE/Open University Press.

— (2012) 'Researching absences and silences in higher education: Data for democratisation'. *Higher Education Research and Development*, 31 (3), 353–68.

Morley, L. and Lussier, K. (2009) 'Intersecting poverty and participation in higher education in Ghana and Tanzania'. *International Studies in Sociology of Education*, 19 (2), 71–85.

Murphy, D. (2006) *Empowerment or Control? A case study exploring the possibilities and limitations of empowerment for undergraduates: Who participated in an Aimhigher Student Ambassador Scheme*. MA diss., London: Institute of Education.

NAE (National Academy of Engineering) (2002) *Raising Public Awareness of Engineering*. Washington, DC: The National Academies Press.

— (2008) *Changing the Conversation: Messages for improving public understanding of engineering*. Washington, DC: The National Academies Press.

Naidoo, R. (2003) 'Repositioning higher education as a global commodity: Opportunities and challenges for future sociology of education work'. *British Journal of Sociology of Education*, 24 (2), 249–59.

National Research Council (2010) *Rising Above the Gathering Storm, Revisited: Rapidly approaching category 5*. Washington, DC: The National Academies Press.

Newburn, T. and Shiner, M. (2006) 'Young people, mentoring and social inclusion'. *Youth Justice*, 6 (1), 23–41.

Newton, P. (1987) 'Who becomes an engineer? Social, psychological antecedents of non-traditional career choice'. In Spencer, A. and Podmore D. (1987) *In a Man's World: Essays on women in male-dominated professions*. London: Tavistock Publications Ltd.

Northwood, M.D., Northwood, D.O. and Northwood, M.G. (2003) 'Problem-based learning (PBL): From the health sciences to engineering to value-added in the workplace'. *Global Journal of Engineering Education*, 7 (2), 157–63.

OECD (Organisation for Economic Co-operation and Development) (2008) *Education at a Glance 2008: OECD indicators*. Paris: OECD.

Office for Public Management, for the Royal Society (2006) *Taking a leading role – scientists survey*. London: The Royal Society.

Ormerod, M.B. and Duckworth, D. (1975) *Pupils' Attitudes to Science*. Slough: NFER.

Osborne, J. and Dillon J. (2007) 'Research on learning in informal contexts: Advancing the field?' *International Journal of Science Education*, 29 (12), 1441–5.

Osborne, J., Simon, S. and Collins, S. (2003) 'Attitudes towards science: A review of the literature and its implications'. *International Journal of Science Education*, 25 (9), 1049–79.

Packard, B. W. (1999) 'A "composite mentor" intervention for women in science'. American Educational Research Association Annual Meeting, Montreal, QC.

Packard, B. W. and Nguyen, D. (2003) 'Science career-related possible selves of adolescent girls: A longitudinal study'. *Journal of Career Development*, 29 (4), 251–63.

Parker, I. (1992) *Discourse Dynamics: Critical analysis for social and individual psychology*. London: Routledge.

Parry, G. (2010) 'Differentiation, competition and policies for widening participation'. In David, M *et al.* (eds) *Improving Learning by Widening Participation in Higher Education*. London: Routledge.

— (2011) 'Mobility and hierarchy in the age of near-universal access'. *Critical Studies in Education*, 52 (2), 135–49.

Pattillo-McCoy, M. (1999) *Black Picket Fences: Privilege and peril among the black middle class*. Chicago: The University of Chicago Press.

Perkins, J. (2013) *Professor John Perkins' Review of Engineering Skills*. London: Department for Business, Innovation and Skills.

Piaget, J. (1966) *The Psychology of Intelligence*. London: Routledge.

Porter, S. (2010) *A Sporting Chance: Boys into higher education using football project*. Report of the evaluation of the second phase. Leicester: Aimhigher Leicester City and Leicestershire.

RAEng (Royal Academy of Engineering) (2007) *Educating Engineers for the Twenty-first Century*. London: Royal Academy of Engineering.

— (2009) *Inspiring Women Engineers*. London: Royal Academy of Engineering.

Ragins, B.R. (2002) 'Understanding diversified mentoring relationships: Definitions, challenges and strategies'. In Clutterbuck, D. and Ragins, B.R. *Mentoring and Diversity: An international perspective*. Oxford: Butterworth-Heinemann.

Ragins, B.R. and Cotton, J.L. (1999) 'Mentor functions and outcomes: A comparison of men and women in formal and informal mentoring relationships'. *Journal of Applied Psychology*, 84 (4), 529–50.

Rainbird, H. (2000) 'Skilling the unskilled: Access to work-based learning and the lifelong learning agenda'. *Journal of Education and Work*, 13 (2), 183–97.

Reay, D. (2001) 'Finding or Losing yourself? Working-class relationships to education'. *Journal of Education Policy*, 16 (4), 333–46.

Reay, D., Crozier, G. and Clayton, J. (2009) '"Strangers in paradise?" Working-class students in elite universities'. *Sociology*, 43 (6), 1103–21.

Reay, D., David, M. and Ball, S. (2005) *Degrees of Choice: Social class, race and gender in higher education*. Stoke-on-Trent: Trentham Books.

Reese, R. (2004) *American Paradox: Young black men*. Durham, NC: Carolina Academic Press.

References

Reiss, M., Hoyles, C., Mujtaba, T., Riazi-Farzad, B., Rodd, M., Simon, S. and Stylianidou, F. (2010) *Understanding Participation Rates in Post-16 Mathematics and Physics: Conceptualising and operationalising the UPMAP Project*. Online. http://eprints.ioe.ac.uk/6731/1/Reiss2011Understanding273.pdf [10/2011]

Renold, E. and Ringrose, J. (2008) 'Regulation and rupture: Mapping tween and teenage girls' resistance to the heterosexual matrix'. *Feminist Theory*, 9 (3), 313–38.

Reynolds, T. (2004) *Caribbean Families, Social Capital and Young People's Diasporic Identities*. London South Bank University, Families and Social Capital ESRC Research Group Working Paper No. 11.

Rhodes, J.E. (2002) *Stand by Me: The risks and rewards of mentoring today's youth*. Cambridge, MA: Harvard University Press.

Ringrose, J. (2007) 'Successful girls? Complicating post-feminist, neo-liberal discourses of educational achievement and gender equality'. *Gender and Education*, 19 (4), 471–89.

Robson, C. (2002) *Real World Research: A resource for social scientists and practitioner–researchers*. 2nd ed. Oxford: Blackwell.

Roth, J.L. and Brooks-Gunn, J. (2003) 'Youth development programs: Risk, prevention and policy'. *Journal of Adolescent Health*, 32, 170–82.

Rubin, H.J. and Rubin, I.S. (1995) *Qualitative Interviewing: The art of hearing data*. London: Sage.

Sainsbury, D., Lord (2007) *The Race to the Top: A review of government's science and innovation policies*. Norwich: HM Treasury. Online. http://www.rsc.org/images/sainsbury_review051007_tcm18-103118.pdf [17/03/2014].

Sanchez, B. and Colon, Y. (2005) 'Race, ethnicity and culture in mentoring relationships'. In DuBois, D.L. and Kercher, M.J. (eds) *Handbook of Youth Mentoring*. Thousand Oaks, CA: Sage.

Sanchez, B. and Reyes, O. (1999) 'Descriptive profile of the mentorship relationships of Latino adolescents'. *Journal of Community Psychology*, 27 (3), 299–302.

Sanchez, B., Reyes, O., Potashner, I. and Singh, J. (2006) 'A qualitative examination of the relationships that play a mentoring function for Mexican American older adolescents'. *Cultural Diversity and Ethnic Minority Psychology*, 12, 615–31.

Sanchez-Jankowski, M. (2003) 'Representation, responsibility and reliability in participant observation'. In May, T. (2003) *Qualitative Research in Action*. London: Sage.

Sanders, C. (2006) 'Tories: £2 billion access drive is a "failure"'. *Times Higher Education Supplement*, 28 July, 2.

Sanders, J. and Higham, L. (2012) *The Role of Higher Education Students in Widening Access, Retention and Success: A literature synthesis of the widening access, student retention and success*. National Programmes Archive. York: Higher Education Academy http://www.heacademy.ac.uk/assets/documents/WP_syntheses/WASRS_Sanders.pdf [17/03/2014].

Schibeci, R.A. (1984) 'Attitudes to science: An update'. *Studies in Science Education*, 11, 26–59.

Schon, D. (1983) *The Reflective Practitioner: How professionals think in action*. New York: Basic Books.

Shiner, M. and Modood, T. (2002) 'Help or hindrance? Higher education and the route to ethnic equality'. *British Journal of Sociology of Education*. Special Issue on the Sociology of Pierre Bourdieu, 25 (4), 473–88.

Siann, G. and Callaghan, M. (2001) 'Choices and barriers: Factors influencing women's choice of higher education in science, engineering and technology'. *Journal of Further and Higher Education*, 25 (1), 85–95.

Silva, E. (2001) '"Squeaky wheels and flat tires": a case study of students as reform participants'. *Forum*, 43 (2), 95–9.

Sipe, C.L. (2002) 'Mentoring programs for adolescents: A research synthesis'. *Journal of Adolescent Health*, 31, 251–60.

Skeggs, B. (1997) *Formations of Class and Gender*. London: Sage.

— (2003) 'Techniques for telling the reflexive self'. In May, T. (ed.) *Qualitative Research in Action*. London: Sage.

Slack, K., Mangan, J., Hughes, A. and Davies, P. (2012) '"Hot", "cold" and "warm" information and higher education decision-making'. *British Journal of Sociology of Education*, 35 (2), 204–23.

Smith, E. and White, P. (2011) 'Who is studying science? The impact of widening participation policies on the social composition of UK undergraduate science programmes'. *Journal of Education Policy*, 26 (5), 677–99.

Spencer, A. and Podmore, D. (1987) *In a Man's World: Essays on women in male-dominated professions*. London: Tavistock Publications Ltd.

Spencer, R. (2007) 'Naturally occurring mentoring relationships involving youth'. In Allen, T.D. and Eby, L.T. (2007) *The Blackwell Handbook of Mentoring: A multiple perspectives approach*. Oxford: Blackwell Publishing.

Starks, F.I. (2002) 'Mentoring at-risk youth: An intervention for academic achievement'. Ph.D. diss., Alliant International University.

Steele, T. (2000) 'Common goods: Beyond the new work ethic to the universe of the imagination'. In Thompson, J. (ed.) *Stretching the Academy*. Leicester: NIACE.

Stirling, M. (2006) Student Ambassador Research in the Aspire Partnership Area.

Strauss, A.L. and Corbin, J. (1990) *Basics of Qualitative Research: Grounded theory procedures and techniques*. Thousand Oaks, CA: Sage.

— (1998) *Basics of Qualitative Research: Techniques and procedures for developing grounded theory*. 2nd ed. Thousand Oaks, CA: Sage.

Straw, S., Hart, R. and Harland, J. (2011) *An Evaluation of the Impact of STEMNET's Services on Pupils and Teachers*. Slough: NFER.

Stuart, M. (2006) '"My friends made all the difference": Getting into and succeeding at university for first generation entrants'. *Journal of Access Policy and Practice*, 3 (2), 162–84.

Sutton Trust (2007) *University Admissions by Individual Schools*. London: Sutton Trust.

Tai, R.H., Qi Liu, C., Maltese A.V. and Fan, X. (2006) *Focus Group Interviews in Education and Psychology*. Thousand Oaks, CA: Sage.

Taylor, S.E. and Lobel, M. (1989) 'Social comparison activity under threat: Downward evaluation and upward contacts'. *Psychological Review*, 96, 569–75.

Taylor, Y. (2008) 'Good students, bad pupils: Constructions of "aspiration", "disadvantage" and social class in undergraduate-led widening participation work'. *Educational Review*, 60 (2), 155–68.

References

Tesser, A. (1986) 'Some effects of self evaluation maintenance on cognition and action'. In Sorrentino, R.M. and Higgins, E.T. (eds) *Handbook of Motivation and Cognition: Foundations of social behaviour*. New York: Guildford Press.

Thomas, L. (2001) 'Power, assumptions and prescriptions: A critique of widening participation policy-making'. *Higher Education Policy*, 14 (4), 361–76.

— (2012) *Building Student Engagement and Belonging in Higher Education at a Time of Change: Final report from the What Works? Student Retention and Success programme*. London: Paul Hamlyn Foundation.

Thompson, J. (2000) *Stretching the Academy*. Leicester: NIACE.

Thompson, M. (2010) *Birmingham and Solihull Associates Scheme Evaluation Report 2009–2010*. Birmingham: Aimhigher Birmingham and Solihull. Online. http://www.heacademy.ac.uk/assets/documents/aim_higher/AHBS-Working_together_to_make_a_difference.pdf [17/03/2014].

Tierney, J. and Grossman, B.B. (2000) 'What works in promoting positive youth development: Mentoring'. In Kluger, M.P., Alexander, G. and Curtis, P.A. (eds) *What Works in Child Welfare*. Washington DC: Child Welfare League of America, 323–26.

UKCES (UK Commission for Employment and Skills) (2009) *Working Futures 2007–17*. Online. http://webarchive.nationalarchives.gov.uk/+/http://www.ukces.org.uk/upload/pdf/Working%20Futures%203%20FINAL%20090220.pdf [17/03/2014].

UNESCO (2009) *Global Education Digest*, 2006. Montreal: UNESCO Institute for Statistics.

Vignoles, A. and Crawford, C. (2010) 'Access, participation and diversity questions in relation to different forms of further and higher education: The importance of prior educational experiences'. In David, M. *et al.* (eds) *Improving Learning by Widening Participation in Higher Education*. London: Routledge.

Vygotsky, L.S. (1978) *Mind in Society: The development of higher psychological processes*. Cambridge, MA: Harvard University Press.

Walker, M. (2001) 'Engineering identities'. *British Journal of Sociology of Education*, 22 (1), 75–89.

Walkerdine, V., Lucey, H. and Melody, J. (2003) 'Subjectivity and the qualitative method'. In May, T. (ed.), *Qualitative Research in Action*. London: Sage.

Watermeyer, R. (2012) 'Confirming the legitimacy of female participation in science, technology, engineering and mathematics (STEM): Evaluation of a UK STEM initiative for girls'. *British Journal of the Sociology of Education*, 33 (5), 679–700.

Watts, M. and Bentley D. (1993) 'Humanizing and feminizing school science: Reviving anthropomorphic and animistic thinking in constructivist science education'. *International Journal of Science Education,* 16, 83–97.

Watts, M. and Bridges, D. (2004) *Whose Aspirations? What Achievement? An investigation of the life and lifestyle aspirations of 16–19 year olds outside the formal system of education*. Cambridge: Association of Universities in the East of England.

Wayne, M. and Meyenn, B. (eds) (2001) *What About the Boys? Issues of masculinity in school*. Buckingham: Open University Press.

General Medical Council (GMC) 19
Gerson, K. 44
Gorard, S. 8, 28
government policy 1–2, 61
graduate premium 149
'grapevine knowledge' 9
Grossman, J.M. 38, 112, 121
grounded theory 55–6
Hager, P. 11–12
Harrison, Matthew 15
Hatt, S. 30, 32, 160
'heterosexual matrix' (Butler) 142–3
Hey, V. 24, 131, 141
Higham, L. 27–9
higher education (HE): expansion in 2–5; marketization of 5; massification of 4; stratified nature of 87; structural divisions within 3–5
Higher Education Funding Council for England (HEFCE) 27–9, 32, 34, 122; advisory group 15–16
Higher Education Statistics Agency (HFSA) 16
Higher Education: Students at the Heart of the System (White Paper, 2011) 1, 5
Higher Education through Football project 32
Hockings, C. 6, 156
Hodkinson, P. 8, 12, 154
Horowitz, R. 44
'hot' sources of information about education 9–10, 30, 74, 80, 87, 112, 153
House of Commons Education Committee 85
House of Commons Select Committee on Disaffected Children 36
Hughes, G. 22–3, 121
identity 22–4; challenging of 146–7; and learning 8–13; matching aspects of 122, 126; performance of 136–42, 157; and widening participation 5–8; *see also* shared learner identity; subject identities
individualism and individualization 6, 81, 87, 115, 151
informal learning 20, 34, 133, 153, 155; attributes of 101, 103
'informed consent' 53
Institute of Education 15, 21
institutional racism 19
intersectionality, concept of 6, 121
Keller, T.E. 35–6
Key Stage 5 qualifications 20
'knowledge economy' 149, 152
Kyriacou, C. 101
Laclau, E. 42
Lammy, David 28
Lauder, H. 149
Lave, J. 11, 114
law, perceptions of 20
league tables 20–1, 33–4, 89, 153

learning: responsibility for 96–7; seen as *acquisition* 8; seen as *becoming* 154; seen as *development of identity* 12–13; seen as *participation* 11; seen as *social activity* 10; *situated* 11
learning practices and identities 88–114, 152–5
Leathwood, C. 4
Liang, B. 38, 112, 121
liberal-humanist tradition in education 5, 42
'licensed mimicry' (McRobbie) 142
Lightbody, P. 22
Lussier, K. 7
Macleod, F. 8, 12, 154
McRobbie, A. 24, 142
Malcolm, Janice 5, 87
marketing 57–66, 70, 86–7, 158; discourses of 61–6, 86–7; of the idea of progression 58; meanings of 149–52; and role models 115–17
marketization of higher education 158
Marx, D.M. 39
masculinity 134, 137–9
Mason, J. 53
mathematics: education in 20–1; exchange value of 94, 113
mature students 9
Medical Access Scheme (MAS) *see* Royal School of Medicine
medical days and medical afternoons 81–6, 101, 104–7, 110–14, 133–4
medical education 15–20
medicine: perceptions of 20; 'selling' of 75–7
mentoring 26–9, 35–8, 95; of behaviour 106; definition of 26
Mentoring Action Project (MAP) 36
mentoring relationships: matching of participants in 37–8, 121; naturally-occurring 37
'Mickey Mouse' courses 4
Mirza, H. 121
modelling of behaviour 117–21, 155
Morley, L. 2, 7, 149
Mouffe, C. 42
Murphy, D. 29
Naidoo, R. 2
National Health Service (NHS) 19
National Mentoring Network 36
National Scholarship Programme 5
neo-liberalism 33, 87, 149–52, 158
Newburn, T. 35
new universities 3–5, 20
Office of Fair Access (OFFA) 4–5
Organisation for Economic Co-operation and Development (OECD) 2
Osborne, J. 15, 20–1
outreach activities 26, 33–4, 46, 48, 88, 100–2, 148, 160–3
Packard, B.W. 38
Paczuska, A. 10, 30

Index

parental influence 17
Parker, I. 41
Parry, G. 5
participant observation 51
Pattillo-McCoy, M. 50
performativity 12, 43, 114, 119, 156–7
physics education 21
polytechnics 19–20
positioning theory 41–3
post-structuralism 12, 22–3, 114
power relations 13
problem-based learning (PBL) 17–18, 101, 153
professionalism 57, 77–9, 87, 91, 150
progression in education 58, 148
promotional activity 58–61, 150–4, 158
quantitative measurement of educational outcomes 1, 8
Quint, J. 37
racism *see* institutional racism
Reay, D. 4, 7, 9–10, 50, 68, 86, 113, 132
regimes of truth 40–2, 66, 86–7, 89, 101, 158
Renold, E. 24
Reynolds, T. 133
Rhodes, J.E. 36
Ringrose, J. 24
Robbins Report (1963) 3
Roberts, T. 16, 18–19
role models 27, 30, 37–8, 56, 110, 115–18, 121, 123, 146, 148, 153, 155, 158; lack of 38; and marketing 115–17
Roman, J.S. 39
'rough' pupils 98–9
Royal Academy of Engineering 19
Royal School of Medicine 3, 44–51, 57–66, 75–87, 95–7, 101–2, 111–13, 115–22, 127, 132–5, 143–4, 147, 149–51, 155–7, 161, 163; discourses of charity and deficit at 80–7; marketing discourses circulating at 61–6; Medical Access Scheme (MAS) 46, 48, 75, 81, 88, 101–2, 117, 150–3, 160–3; 'selling' of medicine at 75–7
Sainsbury, Lord 101
Sanders, C. 30
Sanders, J. 27–9
science, technology, engineering and mathematics (STEM): 162–3; and the economy 14–15; engagement with 20, 24, 38; importance of 151; orientation towards 154–5
'selling' of subjects 69–77
shared learner identity 110–13, 126, 132–3, 154–7
Shiner, M. 35
Siann, G. 21–2
Silva, E. 48
situated learning 114
Skeggs, B. 43, 50, 132, 156

skills shortages 14
Slack, K. 10, 30, 32
small talk 136–41, 157
social exclusion 36
social interaction 11
social learning 9, 12–13
Social Market Foundation 14
social mobility 2–6, 15
social networks 9
social psychology 41
social support for young people 37
socialization between ambassadors and pupils 138–41
socio-cultural backgrounds of pupils and ambassadors 156
special educational need (SEN) 96
Spencer, R. 37
Steele, T. 2–3
STEM *see* science, technology, engineering and mathematics
STEMNET 38–9
Stokoe, E. 42–3
strategically important and vulnerable subjects (SIVSs) 16
Strauss, A. 55
Straw, S. 38–9
Stuart, M. 9, 49
student ambassadors 1, 26–39, 41–2, 46–8, 52–5; age of 136, 139; appearance of 142–3; benefits from work of 27–34; collaborative working with pupils 153–4, 158–9; contribution of 152; definition of 26; different reactions of male and female pupils to 137–9; female 123; future development of schemes for 159; location, contexts and focus for work of 31–2; organizational structures for work of 33–4; physical interactions with pupils 141–6, 157; positioning of 88–100, 104–14, 116, 136–7, 143, 148–50, 153–4; professionalism of 77–8, 91, 150; social identities of 118–33; social relationships with pupils 133–6, 155, 157; used as a marketing tool 57–87
Student Associates scheme 26–9, 32, 34; benefits for the associates themselves 29, 34
student-led activities 106–7
students: choices made by 1, 5–6, 8–10; non-traditional 6
Students into Schools programme 31
subject identities, marketing of 86–7
subjection, theory of 13, 119–20, 157
summer schools 32, 66, 68, 72–6, 79, 98, 101–2, 111, 113, 122, 124, 135, 138, 141–4, 147, 149–50, 161
surface learning 113
Sutton Trust 4
targeting of initiatives 160–2
Taylor, Y. 29–32, 98–9, 153–4
Teacher Development Agency (TDA) 161

183

teaching assistants 96, 100
Train Tracks 47, 72–5, 101, 106–9, 111–14, 122
Tranter, D. 2
T-shirts 58–9
Tube Lines (company) 19
underprivilege 81
United Kingdom Commission for Employment and Skills 14
United Kingdom Resource Centre (UKRC) 21
United States National Academy of Engineering (NAE) 18
universities: differences between 3–5; promotion of (individually and collectively) 58–9, 61
upskilling the workforce 3, 149
Vincent, C. 9
vocational education and training 5, 20, 22

Walker, M. 24, 142–3
Walkerdine, V. 44
Watermeyer, R. 22, 24
Wenger, E. 11, 114
Wetherell, M. 42
widening participation (WP) initiatives 1, 14, 20, 24–5, 42, 57–60, 79, 88, 148–52, 155–8, 160–3; benefits from 26–34; critiques of 9; history of 2–5; and identity 5–8; and research methods 9
Willets, David 7
Williams, J. 20–1, 94, 113
Willig, C. 42, 54–5
Ylonen, A. 29, 32–3
Youdell, D. 119
youth culture 131
Zimmerman, M.A. 37